W9-BDU-456

ALSO BY MANJULA MARTIN

Fruit Trees for Every Garden

Scratch: Writers, Money, and the Art of Making a Living (ed.)

The Last Fire Season

The Last Fire Season

A Personal and Pyronatural History

MANJULA MARTIN

PANTHEON BOOKS · New York

A portion of this work originally appeared, in different form, in the September 30, 2021, issue of *The New Yorker.*

Library of Congress Cataloging-in-Publication Data
Name: Martin, Manjula, author.
Title: The last fire season: a personal and pyronatural history / Manjula Martin.
Description: First edition. New York: Pantheon Books, 2024.
Includes bibliographical references.
Identifiers: LCCN 2023033922 (print). LCCN 2023033923 (ebook).
ISBN 9780593317150 (hardcover). ISBN 9780593317167 (ebook).
Subjects: LCSH: Martin, Manjula. | Wildfires—California, Northern. | Human beings—Effect of environment on. | Women authors, American—Biography. | California, Northern—Biography.
Classification: LCC SD421.32.C2 M37 2024 (print) | LCC SD421.32.C2 (ebook) | DDC 363.37/909794—dc23/eng/20230825
LC record available at https://lccn.loc.gov/2023033922
LC ebook record available at https://lccn.loc.gov/2023033923

www.pantheonbooks.com

Jacket photograph by George Sherman
Jacket design by Jenny Carrow
Maps copyright © 2024 by David Lindroth, Inc.
Interior glyph designs adapted from tattoo art by Jessica Zed

Printed in the United States of America
First Edition
2 4 6 8 9 7 5 3 1

For Max, obviously

If we opened people up, we'd find landscapes.

—AGNÈS VARDA

Contents

OCTOBER

NOVEMBER

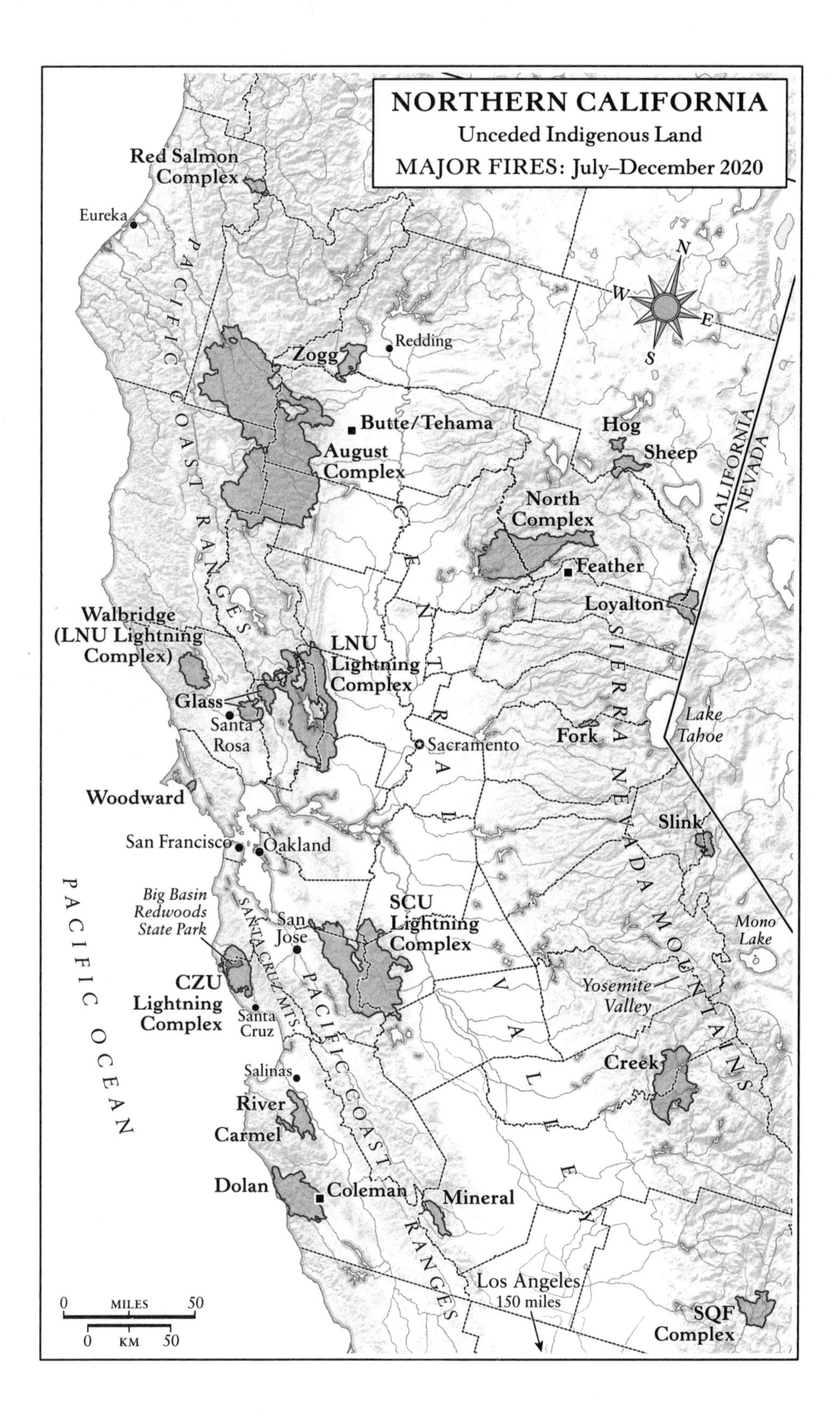
NORTHERN CALIFORNIA
Unceded Indigenous Land
MAJOR FIRES: July–December 2020
Red Salmon Complex
Eureka
Redding
Zogg
Butte/Tehama
August Complex
Hog
Sheep
CALIFORNIA
NEVADA
North Complex
Feather
Loyalton
PACIFIC COAST RANGES
CENTRAL VALLEY
SIERRA NEVADA MOUNTAINS
Walbridge (LNU Lightning Complex)
LNU Lightning Complex
Glass
Santa Rosa
Sacramento
Fork
Lake Tahoe
Woodward
San Francisco
Oakland
Slink
Big Basin Redwoods State Park
SANTA CRUZ MTS.
San Jose
SCU Lightning Complex
Mono Lake
CZU Lightning Complex
Santa Cruz
Yosemite Valley
PACIFIC OCEAN
Salinas
Creek
River
Carmel
Dolan
Coleman
Mineral
Los Angeles 150 miles
0 MILES 50
0 KM 50
SQF Complex
N
W
E
S

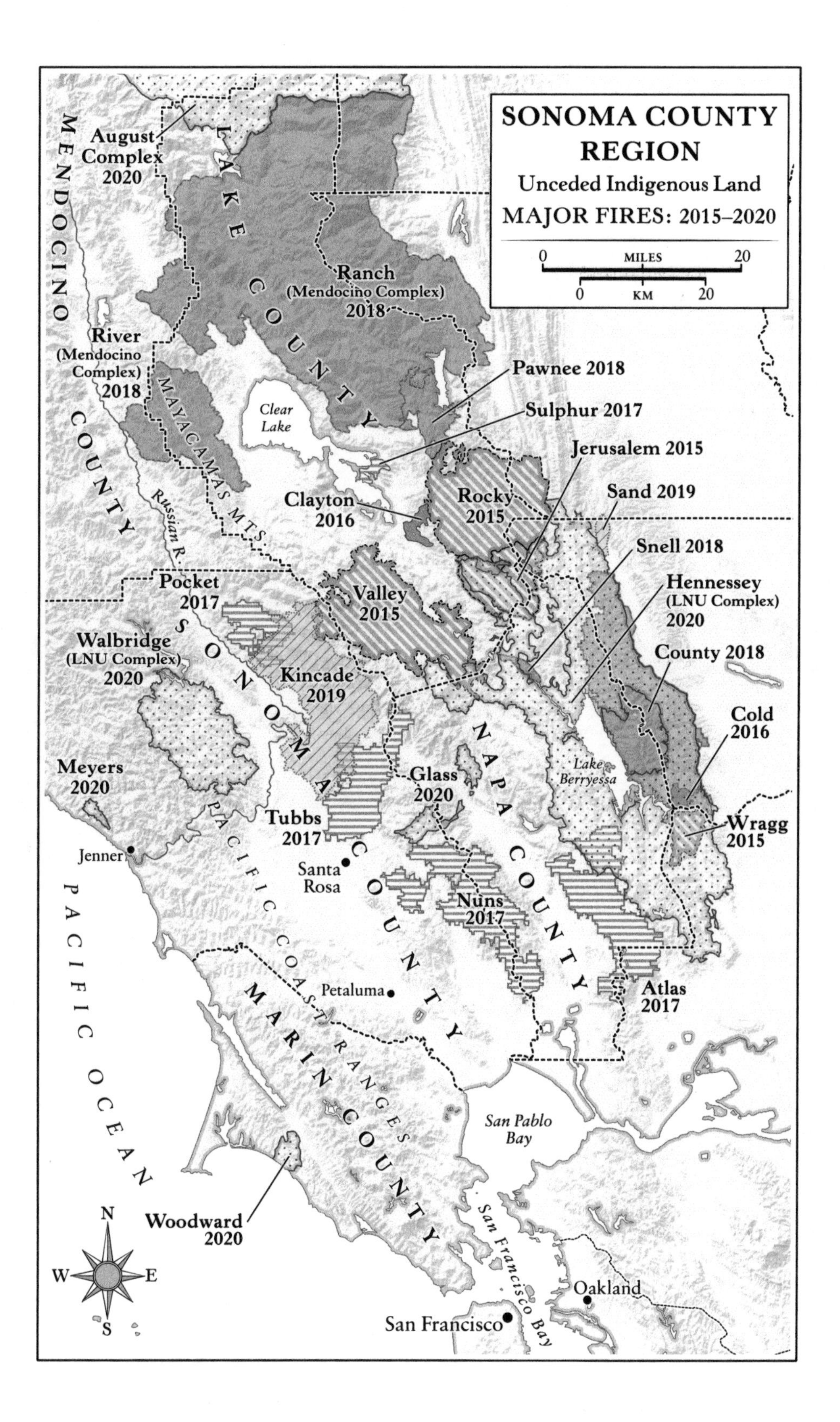
SONOMA COUNTY REGION
Unceded Indigenous Land
MAJOR FIRES: 2015–2020
0 MILES 20
0 KM 20
August Complex 2020
Ranch (Mendocino Complex) 2018
River (Mendocino Complex) 2018
Pawnee 2018
Sulphur 2017
Jerusalem 2015
Clayton 2016
Rocky 2015
Sand 2019
Snell 2018
Hennessey (LNU Complex) 2020
County 2018
Cold 2016
Wragg 2015
Pocket 2017
Valley 2015
Walbridge (LNU Complex) 2020
Kincade 2019
Meyers 2020
Glass 2020
Tubbs 2017
Nuns 2017
Atlas 2017
Woodward 2020
MENDOCINO COUNTY
LAKE COUNTY
SONOMA COUNTY
NAPA COUNTY
MARIN COUNTY
MAYACAMAS MTS.
PACIFIC COAST RANGES
PACIFIC OCEAN
Clear Lake
Russian R.
Lake Berryessa
San Pablo Bay
San Francisco Bay
Jenner
Santa Rosa
Petaluma
Oakland
San Francisco
N
S
E
W

August

· 1 ·

Storm

The night of a dry lightning storm in Northern California I woke up terrified, and from my bedroom window I watched relentless spears of lightning shatter the sky, Zeus or Jupiter very upset, fire from darkness splintering the land, and I knew immediately we couldn't all survive this. People. Critters. Houses. Trees. Max rolled over in bed next to me. There's going to be a fire, I said. We got up. It was four-thirty A.M.

I lived on the north slope of a thickly forested hill with Max, my partner, in a small white house under large red trees. Our road was a single-lane dead-end. Down the hill there was a ramshackle neighborhood of cabins. Upslope, past the leaning fence of our yard, there grew three hundred acres of mixed evergreen forest, which was privately owned and chronically neglected. From here the woods spread west into thousands more acres of Northern California landscape: more trees, more hills, small groupings of homes perched between. Farther west, where the coastal range stopped, shrubby slopes descended to the Pacific Ocean. My extended backyard.

Wind had woken me before the lightning; it rattled the single-pane windows in our bedroom. Above the redwoods fathomless clouds lingered like silence. From inside them the furious sky hurled its energy at millions of acres of dry, deep wood. I had never seen so many lightning strikes. The blades of electricity bisected the air, the earth, everything. My insides were set abuzz. My lungs contracted like they'd just hit cold water; my jaw compacted into itself; my eyes searched for purchase in the uneasy dark. Every muscle in my pelvis, from psoas to sphincter, felt as though it had been turned to wood. Somewhere inside my brain every synapse fired, and I was thrust into a whorl of anxiety: go, go, go.

The storm continued. Max and I ping-ponged between each window in our house, trying to track the lightning and gauge its proximity to the roof; the large, open yard; the 150-foot-tall redwood trees surrounding it; the thousands more trees in the hills. We opened the door and stood on the back porch beneath the eaves and looked up. The canopy blocked our view of the dive-bombing sky.

Redwoods were the tallest plants on the planet, older than almost everything. Since childhood I had felt safe beneath the shelter of these grand trees. I often thought of them as my protectors, and myself as their comrade. The redwoods where I lived—coast redwoods, *Sequoia sempervirens*—were second- and third-growth, as most in the area were, due to past logging. They were probably over a hundred years old; just babies, in redwood years. The trees lived together with us in mutual silence, and when it was windy they swayed gracefully above the roses I'd planted in the clearings between them, as though they were keeping watch. Despite my feelings of comradeship, past storms and human history showed that the trees and I were in fact liabilities to each other, not guardians; anyway, we couldn't protect each other from this. Lightning was inescapable, an elemental force unleashed. It struck and struck, splintered and shone. My skin bristled as the atmospheric pressure plummeted, but bizarrely this lightning had arrived without rain. The storm was near to us, very near, and every time the thunder clapped I counted one, two, three inside my head to clock its proximity. But the expected crescendo of every thunderstorm, the deluge, never came. Instead, electric spears kept plunging toward the earth and fear kept rising inside my body, and the two connected in my brain and, perhaps, never came untied.

We went back inside. Max looked online for reports of new fires, and I put on a sports bra, in case I had to run from something. I then walked to the closet where we kept our camping gear and started to take stuff out. Although we'd evacuated from a wildfire the prior year, we didn't have an emergency kit—a *go bag*, in disaster-preparedness parlance, which was fast becoming everyday lingo all over the world. Here, fires didn't usually happen until autumn. In recent years I had noticed less predictability to the seasons, but by August the land was reliably dry, the hills a mélange of browns and yellows. August nights, however, were moist; in the mornings the fog crept out of the valleys and back toward the ocean like it was hungover. By September the marine layer would relax its grip as the Diablo winds, named in part for their capacity to do bad things, began to roll over the mountains from the east, and the oak leaves and fescue would then shimmer like hot gold. Historically, in summer in coastal Northern California, it did not rain. It did not thunder. Lightning was for other seasons. But there was so much fire in that sky. It had to land somewhere.

I did laps around the house, carrying things from closet to bed. Tote bag, sleeping bag, head lamp. Car registration, my asthma inhaler, passports. In the closet we had an old, cheap backpack with a broken strap that had been my feeble earthquake kit when I had lived alone in San Francisco. I found it, cursed its uselessness, and tossed it on the floor near the front door anyway. I had read that in the extreme heat of a wildfire scenario, synthetic clothing could fuse to your skin, whereas natural fibers burned clean. I positioned my leather clogs by the door and fingered the fabric of my pajamas. Cotton. I added to the pile a wool sweater, in case it ever got cold again, although we'd been in a severe heat wave all week. The lightning kept crashing and the electricity flickered, but held. Max reported back from the internet: no fires yet. Wild lightning pics, though. I kept packing. I thought I could smell sulfur. Somewhere in my body, a habitual response had already taken control. The world was flexing its power over me, and I knew from experience that when this happened it was important to be quick, be ready, then be gone.

At some point our local firehouse siren moaned its air-raid lament. I checked my phone. I checked again. No notifications. But the siren meant that someone nearby had called 911, which meant that somewhere near my home, trees were being cracked open; power lines were falling; small fires were certainly starting. We could only hope that the volunteer fire department was finding them all. Our next-door neighbor was on the VFD; I usually relied on him for intel about storms or fires. I looked out my kitchen window and across the half-acre slope of the yard toward his driveway, but his pickup truck was already gone.

Max joined me in packing. We didn't talk much. My thunder counts grew shorter in number, then began again to extend. The lightning stopped, eventually. The sirens quieted. Then the light dawned and it was gray, then, bizarrely, there was a very small amount of rain, more thunder, and for a refreshing moment the atmosphere felt like August in Pennsylvania that summer when I was fifteen years old and away from California for the first time and the thunder rolled across the green landscape every afternoon and two different boys wanted to kiss me. But this moisture was anomalous, limited, I knew. It couldn't make magic happen.

It was 2020, the first year of the coronavirus pandemic. There were no vaccines. On Earth's surface it was about one degree Fahrenheit warmer, on average, than it had been when I was a child. In the United States a climate denier with white-supremacist leanings occupied the highest political office. Around the world people and places were having a hard time. It was a year of overwhelm: In January flash floods had displaced sixty thousand households in Indonesia. I had been told to stop hugging my friends in late February, and we'd stopped going inside one another's houses in March. That April unseasonable wildfires had burned in northern climates such as Scotland, Poland, and in Ukraine inside the exclusion zone at the site of the Chernobyl nuclear disaster. By May the virus had killed one hundred thousand people in the United States alone. In June a wave of protests swept the country in response to ongoing murders of Black people by the police. The U.S. mail stopped arriving promptly sometime in July. In early August an inland hurricane called a derecho had decimated parts of the Midwest, causing $11 billion in damage and destroying nearly seven million trees in Iowa, and it had barely made national headlines. By this time in the age of humans, it was a known truth that extreme weather events were linked to irreversible changes in the climate—which were caused by increased greenhouse gases in Earth's atmosphere, a direct effect of burning fossil fuels and a slightly less direct effect of capitalism—and that nobody with power seemed to be doing anything differently.

In the San Francisco Bay Area, then home to 7.7 million people, it felt to many like the summer season had seen higher temperatures than usual. The rents were up, too: average monthly rent for a one-bedroom apartment was $3,000 and rising; there were roughly twenty-eight thousand people living without houses in the region. In Sonoma County, two hours north of San Francisco, my mortgage was $2,800 a month. Max, a union organizer, and I, a literary magazine editor, worked from home now. Behind the house in which we lived, the forest trails, known only to locals, had lately on weekends become as crowded as the beach. One neighbor had taken to patrolling the paths for interlopers, mad with territorial defensiveness. There were rumors of increased crimes. People were out of work. People were bored. The air was thick with worry. In many ways, a freak lightning storm fit right in.

That Sunday morning of the storm, after it became light, Max and I checked news and fire maps online but didn't see anything notable. We put on our jauntiest cloth masks and went for bagels at our favorite outdoor café. We made small talk about the eerie weather with the proprietors, a cheery couple in their thirties who cured their own lox, and traded rumors with café regulars about small, lightning-adjacent catastrophes—spot fires that were easily extinguished, power fritzes that were soon restored. We returned home to work on our go bag, a term I disliked because it was grammatically awkward, but I also disliked the alternate, *bug-out bag,* which made me think of doomsday preppers, who in the year 2020 in America were often also racist separatists and whom, as a fellow white person, I felt it important to not share vocabulary with. We spent the day packing the bag without the appropriate moniker: first-aid kit, camping equipment, cash, two kinds of masks (cloth masks, which at the time were recommended for Covid-19, and N95 masks, for smoke), and a medical handbook called *Where There Is No Doctor* that an anarchist friend gave me a decade back. We joked that we weren't sure whether we were packing for a potential wildfire evacuation or an apocalypse, and really don't they go together, one after another?

The next day, a Monday, I flaked off work and drove around doing errands. In the nearest big town, about twenty minutes east, I got more

Band-Aids and refilled my pain meds, for the go bag. On the way home I sped west down Occidental Road with the radio up and the windows down in the oddly humid heat, singing along to David Bowie's "Five Years." As Bowie and I hit the high notes, I felt a peculiar sense of triumph, as though I had done something to merit the luck of not having my home catch on fire. By the time I got home Max had made dinner, which we ate on the deck. When no large fires made their presence known to us that evening, we left the go bag by the front door and went to bed. The air was still sticky and stormy. There was more lightning in the region overnight, but if it came to my house, I didn't wake for it.

The wildfires that came to be known as the 2020 Lightning Complex fires started during these unusual August storms, but the lightning had struck deep in unpeopled places and so humans didn't really know about them for another day or so. As I ran errands and felt alive, the fires spread.

In his book *The Pyrocene,* fire historian Stephen J. Pyne wrote that because fire moved through biomass and consumed life like it was hungry, fire—much like a virus—was often attributed the qualities of a living thing. The language widely used to describe fire further personified it as a thing possessed of violent appetites: fire *devoured* and *raged*. But, Pyne wrote, fire was not a thing; it was a reaction. The science of fire was at its core a simple equation: when fuel, heat, and oxygen met a source of ignition, fire happened. There were generally two entities that caused fires: people (who in California in the twenty-first century were responsible for 95 percent of all wildfires), and lightning (5 percent). Although technically lava and other geologic reactions could start fires, too, lightning had been the primary source of fire on the planet named Earth since it began. Lightning, with its force and unstoppability, was what made fire so powerful. It was the impetus of a raw element sent from the heavens, fire and brimstone, older than time.

When lightning hit a tree, it might cleave the primary trunk open like a wound and simmer inside it before shooting up the tree's limbs and out into the air seeking further fuel. Or, the force of the strike might simply knock flaming bits of wood into the passing wind. Either way, the tree was transformed into a torch. Over seventy-two hours that stormy August, more than ten thousand individual lightning strikes occurred over a landscape that was overgrown, dry from drought, and experiencing record-breaking heat and high winds. The resulting 650 wildfires that were big and serious enough to be named by the fire suppression authorities were simply a matter of odds, earth science at work. But they seemed to behave more like something out of myth.

Northern California was a very large place. When fire came to the northeastern side of the Sacramento Valley at the edge of the Sierra Crest—about a five-hour drive from my house—it delivered upon the watersheds and canyons of Butte and Plumas Counties twenty-one sizable conflagrations that came together to be called the North Complex. There the fire fed on ponderosa pines, toyon, and blue oaks, in the process killing deer and bears and sixteen humans. Up in the emerald forests of Mendocino County, the August Complex fires began their ravenous progress toward the burning of what would become a million acres of trees, consuming the habitats of an untold number of spotted owls in the process. Near Silicon Valley, south and east of the city of San Jose, twenty fires that joined as the SCU Complex vaulted from grass to manzanita shrubs to oak trees, scattering in their wake feral pigs, mountain lions, and red-legged frogs. Between that valley and the ocean lay a coastal mountain range where forty-three years earlier my mother had given birth to me in a trailer next to a geodesic dome. In those same mountains, fire flew with vigor through groves of thousand-year-old redwoods, up to pine ridges, then back downhill across coastal scrub and farmlands, all the way to the sparkling Pacific Ocean. All across the northern half of California, fire forded rivers and freeways. It surrounded small towns and lakes. It barreled through the biomass of the coast ranges, the Sierra foothills, the suburbs along the I-80 corridor, and toward the ruins of a town named Paradise, again. It feasted on woods that adjoined wetlands atop the San Andreas Fault. And in Sonoma County and neighboring Napa County, locus of many prior burnings, fire came to the vineyards and dry valleys, to live oaks and dead tan oaks, guzzling up oily bay trees and unkillable broom, leaping across shingle roofs and over hot hills and tree crowns, through temperate rainforests, and into the woods a few miles north of where Max and I slept.

The season had begun.

On Tuesday morning, two days after the lightning woke me up, it began to rain ash. The sky overhead was blue, the air not smoky, but still the ash came in a slow downward drift. I knew there had been a few holdover fires from the storm, but I hadn't yet checked on their location or progress. I was outside watering the roses when I saw a small black object on the brown grass. I leaned to pick it up. It was a single California bay laurel leaf, and it was burned black. The slender appendage had been sent by the wind like a message over many miles, straight from the heart of a lightning fire to my hands: this dead thing. I held it with caution. In mixed forest habitats like the one I lived in, bay trees were notorious fire spreaders. If left unmanaged, they grew tall and fast to compete for sun, and in a wildfire scenario served as ladder fuels that allowed fire to climb from the forest floor into the canopy, where it could move uncontrolled and kill even large, fire-resilient trees. Bays also acted as a host for sudden oak death, the disease that had killed more than fifty million oak trees in California since the 1990s. I soon found another burned leaf near the apple trees, then another. Bay leaves were redolent of cooking, their scent summoning sense memories of soup in wintertime. But these ones smelled like bad things. I took my ominous treasures inside and on the coffee table created a bleak rainbow of five bay leaves in various shades of burnt, from lightly singed to crispy black. Max was in the other room, on a video call for work. I was technically at work, too. I opened my computer and pulled up the satellite fire map. There they were: the

Walbridge Fire a few miles north of us, the Stewart Fire a few miles to the northwest on the coast near Jenner, the Hennessey to the east, and the other 647 named conflagrations, represented on the map as a spectrum of yellow and orange and red dots forming a perfect circle around my home. Paralyzed, I closed the laptop.

In the early 2000s, sociologist Kari Marie Norgaard studied a Norwegian town that had been severely impacted by climate change via a dearth of snow. She found that, even when met with irrefutable evidence that their environment was transformed for the worse, many residents appeared to ignore what was happening. They wouldn't talk about it. It wasn't a lack of emotion that silenced them; it was the intensity of their feelings. "The word ignore is a verb," Norgaard wrote in her book *Living in Denial: Climate Change, Emotions, and Everyday Life.* Although ignorance might appear to be passive, Norgaard wrote, in the case of climate change such behavior is the product of a complex, systemic social reaction that allows humans, in the face of potential extinction, to downplay our feelings so that we can continue to go about living our lives. In order to conform to social or community norms people might ignore massive events like climate change, even when faced with blatant and unavoidable evidence of the event. They might modulate the way they talked about such events in order to uphold their own internal narratives about themselves or avoid talking about it at all in order to assuage feelings of guilt. Or, they might simply disregard reality as a way to cope, to keep going. Sociologists called these behaviors "emotion management." After the ash started falling, I managed my emotions by going about my day. I worked. I took breaks. I checked the maps online every few minutes as they grew into family trees, crowded with new names and branches of fire offspring.

There seemed to me to be some whimsy to how a life-changing event for thousands of people was named, but the convention was to name a fire after a geographical feature or road near its starting point. The naming of a wildfire was often done by the emergency dispatcher who initially got the call, or by the first firefighting unit on the scene. When geographic details were unknown or sparse, as in very remote areas, a fire sometimes initially received a number, based on its coordinates on the map. The Walbridge Fire, currently burning closest to me, was named for a nearby ridge; it had first been called the 13-4 fire. If a fire continued to burn after its initial discovery, Cal Fire—the statewide agency whose own full name was the California Department of Forestry and Fire Protection, although the forestry part had long ago lost precedence—was responsible for codifying the fire's name. Sometimes names changed as fires merged or grew. In nearby Napa County, the 14-4 fire became the Hennessey, named for a lake, which in turn combined with the Green, Gamble, Spanish, Markley, and Morgan fires, to form just one branch of the LNU Lightning Complex. In situations where there were many fires at once, such as lightning storms, Cal Fire grouped individual fires into complexes. A complex fire could be several fires that had merged together, or several separate fires in a similar area; the name simply corresponded with the command unit of Cal Fire that was responsible for that particular firefight. LNU was the abbreviation for the Cal Fire unit that covered Lake County, Napa County, and Sonoma County (although Sonoma's initial had somehow been omitted). Fire names could sound flippant—Tango, Witch, Bootleg—or evidence the mood of a firefighting team, as in the famously wry Not Creative Fire in Montana in 2015. The Camp Fire that burned down the town of Paradise in 2018 was nothing like a campfire; it was velocious and deadly.

In the afternoon I finished my day-job emails and switched into writing mode. I was trying to make progress on a novel, but instead of my own words, the names of all the fires, all the numbers and streets and command units, overloaded my brain. It was impossible to work. Soon, outside my head, I was interrupted by a roaring noise.

Across the road my neighbor Frank was blowing redwood duff off his roof with a leaf blower powered by gasoline. He stood atop his house shirtless—he was always shirtless—in flip-flops and shorts, his orange-tan skin folded like leather over a thin waist, and he yelled down at me: Ashy!

Frank was in his early seventies and had lived in the neighborhood since 1972, when he got out of three back-to-back tours of duty in Vietnam. He had vacationed here as a child; after he was discharged, he came directly back and never left. Frank loved sunshine, Diet A&W

Root Beer, and his wife, Celia, who kept a lovely potted garden in front of their home. I tilted my head up to him and asked, Aren't you guys packing? Seems like there'll be evacuations.

Frank vroomed the leaf blower and shouted over it at me. We're good, he said. The fire's across the river.

Frank, it's really close, I said. I thought of my bay leaves. I yelled up at him, I think I'm gonna pack up, just in case.

He skittered across the spine of his roof, still denying possibility. Redwoods don't burn, he said.

Maybe he was managing his emotions, too. To Frank, our neighborhood was the world. I prodded him a bit more between bouts of leaf blowing, but he kept saying it can't cross the river, never gonna happen, they'll push it back north.

Frank was wrong about redwoods. They did burn. There were redwoods burning right now across the river. Coast redwoods only grew along a stretch of coastline from Big Sur in the south to the southwest corner of Oregon. They could live to be more than two thousand years old and grow more than three hundred feet tall. The trees thrived in, and created, damp forest environments, which were by definition less fire-prone. Like all trees, redwoods absorbed water from the soil, but they could also literally drink condensation from the sky through foliar uptake. They were dependent on coastal fog for up to 40 percent of their water. Coast redwoods were uniquely fire-resilient in part because they were so tall; their height made fire less likely to reach the crown. Nearer to the ground, foot-thick bark essentially acted as a shield. The bark was high in tannins, the naturally flame-resistant chemical that also gave the trees their red coloring, and low in saps and other fire accelerants. In Northern California, firefighting units based in redwood country were sometimes called asbestos units because it was assumed redwood forests were fireproof. Even if fire got inside a redwood the tree could survive, as evidenced by the common sight of burned-out triangles, often called cat eyes or goose pens, at the base of older trees. Over millions of years of evolution, redwoods developed the ability to recover after crown fires by resprouting epicormically—sending out new shoots directly from their trunks—in addition to the more common survival tactic of basal resprouting, in which a tree sent up new shoots from its existing root ball. Technically a cluster of redwoods was one tree; their shallow roots connected underground, supporting one another and, through mycorrhizal networks in the soil, exchanging nutrients, water, and information. If a mother tree was killed, dozens of basal sprouts would rapidly arise around its footprint, encircling the larger stump of the original tree and creating a younger grove of interconnected redwoods. Such configurations of trees, where I came from, were called fairy rings. Fanciful as the term was, it expressed a truthful sense of how magical these trees could seem. Redwoods could, and did, catch fire. They were just hard to kill. But the people and houses beneath them were thin-skinned. A redwood might survive a fire, but that didn't mean I would.

The neighbors next door to Frank were also prepping their house for potential embers. Jay, a white man around eighty years old wearing a jaunty newsie cap, chatted amiably with me while his younger husband, Richard, ran around laying out hoses and covering their woodpile with an aluminum-foil safety blanket. Richard had been deeply spooked by the catastrophic 2017 fires in Sonoma County and had founded the neighborhood fire-safety group. He wasn't one to project false optimism, but he was saying it, too—the fire has never crossed the river—they were all saying it, over and over, as though it were a prayer: Never happened. Unprecedented. Redwoods don't burn.

I started to feel panicked. I nodded and said something about wind patterns being favorable. Then I began to pack my car.

Three years earlier it had been unanticipated when I, a tattooed, diehard city girl, moved to the woods and started wearing clogs instead of combat boots. It had been unprecedented when, on the night of our housewarming party that same year, a firestorm driven by ninety-mile-an-hour winds burned down several neighborhoods in Santa Rosa, the city ten miles to the east, before firefighters could even arrive, not that they could stop it once they did. It had been unprecedented that the following year, in the space of a couple of hours, a fire would incinerate a town that was home to eighteen thousand people, and that the town would be named Paradise, and the smoke of its dead would cling like fog to the sylvan hill on which I lived, more than one hundred miles away, for an entire month. It had also been extraordinary when, the next year, another wildfire forced most of Sonoma County to evacuate all at once in the middle of the night. You just couldn't know anything anymore. Life wasn't like it had been in the seventies, when the fog's schedule was as reliable as the old hi-fis that Frank tinkered with on his porch. I didn't know how to explain to him that there was no handicapping the current heat of the human race. One simply had to run.

When fire first came to the land now called California, it came to stay. It came as lightning, magma, mountains. Fire was always a naturally occurring part of the landscape in the western United States. From forests to grassland to desert, the diverse collection of ecosystems that made up the vast region called California evolved with fire. The land depended on the cycles of renewal fire brought, and the people who had lived on the land since time immemorial learned to manage this place with an understanding that fire would always need to be a part of it. The relationship between fire, land, and people was interdependent and continually evolving. While pre-colonial Indigenous land management practices did have a disruptive effect on ecosystems—all human presence did—they did so in a way that achieved balance between individuals and community, people and nonpeople. Fire cleared dangerous fuels from the landscape, preventing more-destructive fires. It helped hunters direct the course of game. Fire encouraged the health and frequent regrowth of plants necessary to survival for people and animals: young acorns made staple foods; newly resprouted reeds and grasses made baskets and boats. The specifics of pre-colonial Indigenous land management and cultural practices differed between communities, but fire figured prominently in most of them. It was a primordial physical and spiritual presence—an element, essential to all life.

When new people came to this land in the 1500s, they called themselves Europeans, Spanish, and later Californios. They kept coming, on and off, for hundreds of years, as did people who called themselves Mexican, Russian, or American. And they brought with them their god. During the Mission era, from the mid-1700s to the mid-1800s, California's Indigenous people were displaced, forcibly converted, and enslaved under the missionary systems of settlement carried out largely by Catholic priests (*padres*, in Spanish)—a system explicitly designed to remove people from their land and convert them to a productive peasant population. Then somebody struck gold, and the floodgates of settlement opened even wider.

Beginning with the missions, colonialists had recognized the importance of fire to Indigenous life. In 1793 the acting governor of California under Spanish rule, Don José Joaquín de Arrillaga, issued an edict outlawing the use of fire by Native peoples, even in remote areas, in order to preserve pastureland for Spanish cattle concerns. By 1821, possession of California passed from Spain to Mexico. Alta California, as the province was then known, relied increasingly on livestock economies; fire was used by some landowning rancheros to clear pasture land, but aboriginal fire use was still suppressed. In addition to being important to Indigenous people for ecological and cultural reasons, fire was a tool of war and as such might be used in rebellion. Fire was strategically used in Indigenous uprisings against colonialization in 1776 at Mission San Luis Obispo, when rebels set fire to the roofs of the compound, and in the 1824 Chumash revolt, which led to the occupation of three Southern California missions for several months before being violently quashed by Mexican military forces. Two years later, in Sonoma County, a Native resistance movement—likely including Wappo, Coast Miwok, Suisin, and Patwin peoples—made use of fire by burning several buildings in the mission compound there, ultimately ousting the violently cruel padre in charge.

By the 1850s, California was in the process of becoming a U.S. state after being bounced between colonizers for hundreds of years. One of the first things the California legislature did was create the ironically titled 1850 Act for the Government and Protection of Indians, also known as the Indian Indenture Act. The law prohibited the use of intentional fire in the new state. It also effectively legalized the kidnapping and enslavement of Native people, including the sex trafficking of girls and women. This law and many others like it were part of a coordinated campaign to exterminate—that was the word the then governor used—the Indigenous population of the new state. The campaign was so blatantly intentional that most local governments offered cash bounties to settlers for the body parts of Native people, as proof of their murder. The Indian Indenture Act was revised in 1860 to expand the allowance of enforced labor, permitting Native children to remain enslaved well into adulthood. Over the next twenty years federal, state, and local governments organized or enabled the killing of 80 percent of the Indigenous people then living in California, an act of genocide that among other things erased or pushed underground millennia of traditional ecological knowledge, including the practice and culture of fire. Despite the prohibitions, Indigenous survivors of the genocide continued to use fire on the land when they were able, as did their present-day descendants.

In the late 1800s, conservationists fell in love with California's epic landscapes and sought to preserve them for the use of the public, by which they meant mostly people like themselves. It was assumed or decided that the well-tended forests in places like the Sequoia Forest (the second national park in the United States, after Yellowstone) or the Yosemite Valley (the third) were naturally occurring, and not the result of careful land management over millennia. This view of nature—as a thing separate from man, a museum of the outdoors empty of occupants and frozen in time—combined with the lie that Indigenous people were unsophisticated in their ways of living, thus cementing the American mythology of the West as a place that was both wild and virginal.

Settler culture and old-school conservation culture met in a fear of fire. Since the 1800s some U.S. forestry professionals, usually those with firsthand field experience, had advocated for forest management policies that emulated long-standing aboriginal practices. Because of a timeless combination of racism, profit motives, and the prioritization of recreation and housing over the health of the landscape, they lost. The U.S. Forest Service was established in 1905, with fire prevention and suppression as primary goals. Total fire exclusion was further codified after the Great Fire of 1910, nicknamed the Big Burn or the

Big Blowup, a wildfire driven by extreme winds that killed eighty-nine people and torched three million acres in Montana, Washington, and Idaho over the course of two days. The Big Burn further swayed public opinion against the idea that a fire might be capable of doing anything but causing disaster. By 1935, the U.S. Forest Service had instituted the ten A.M. policy, which stated that every new wildfire must be extinguished by ten o'clock the next morning. The policy was in place until the mid-1970s. In practice, some non-Indigenous landowners in California were able to use prescribed burning without legal consequence well into the 1940s and 1950s, until popular opinion turned with the help of Smokey Bear, the U.S. Forest Service's threateningly peppy anti-fire mascot. It wasn't until the late 1960s that some forestry management entities began to officially reintroduce the practice of controlled burning. By then, the seeds of the current cataclysm had already germinated.

Since the days of colonization and extraction, a deadly mix of factors created the wildfire crisis. The genocide led to the erasure and criminalization of traditional ecological knowledge and practices; it also, in combination with settler conservation philosophies, enforced a culture of fear around fire, further evangelizing the ideology that nature was outside of and subject to humanity. Total fire exclusion and rampant resource extraction led to overcrowded and unhealthy forests that, without regular fire regimes, encroached on open grasslands, creating more heavy fuel. At the same time, California's perpetual cycles of boom/bust development and housing crises continued to send people of all economic classes into the wildland-urban interface (known as the WUI and jarringly pronounced *woo-ee*), where built environments touched up against so-called wild places. I was one of those people. The increased human presence in the WUI primed landscapes with even more fuel, in the form of houses and businesses. And the big one: human-caused climate change from the burning of fossil fuels caused an increase in extreme weather, pushing California's diverse ecosystems—which were already adapted to regular cycles of drought, flood, and fire—toward new polarities. Winter weather oscillated more wildly between record drought and record rainfall. More rain made more vegetation, while hotter dry seasons transformed all that moist, vigorous plant growth into dry kindling. Abundant fuel stoked bigger, hotter, and faster fires. At the same time, increasingly severe winds helped turn regularly occurring wildfires into megafires; more destructive wildfires further decimated ecosystems and property. It was a perfect storm of causation.

California's ecosystems—like those in many parts of the Americas, the Mediterranean, and Australia, to name just a few—were fire-adapted. The land here had been accustomed to regular fire and hadn't been getting it. But even in a fire-adapted landscape, not all fire was good. These new extreme, drought- and wind-fueled wildfires could permanently alter ecosystems. Fire, too, was changing. Wildfire was now occurring in environments entirely unaccustomed to it, such as Siberia, Sweden, and Hawaii. The wildfire crisis, like climate change, was global. The megafire future was now.

Sonoma County could have been a poster child for California's particular variety of the crisis. The place that I called home had been for more than ten thousand years the home of the Coast Miwok, Kashia Pomo, Southern Pomo, Patwin, and Wappo peoples. They were just a few of the many politically, geographically, and linguistically autonomous nations across the American West whose land management practices included intentional, controlled fire. When Russian trappers and European loggers came colonizing here in the 1800s, trees became timber—a resource to be extracted and capitalized—and fire became a crime.

In 1866, in the redwood forests of western Sonoma County, a white carpenter from New Jersey came to seek his fortune. He set up camp in a patch of forest a few miles south of the Russian River and inland from the Pacific Ocean, with the aim of becoming a timber baron. It was, and remained, Coast Miwok and Southern Pomo territory. He logged the redwoods and Douglas firs, became a power player in the nearby town, and cleared space on a steep slope above a creek for more than two hundred vacation cabins. The newly minted logger baron began marketing the land as a camp, a summer-cabin enclave for San Francisco residents, who could take a brisk four-hour train ride to an idyllic forest setting that, as a newspaper ad for the development boasted, "favorably compares with Switzerland and our own Yosemite." In truth, the sylvan hill was only about one thousand feet high (Yosemite's peaks were closer to ten thousand) and the view faced north away from the sun, but the trees were majestic. And the cabins were cheap: $20 each. By the 1950s, when my neighbor Frank's parents bought one, the price had risen to $100. (In 1959 the average cost of a house in California was around $12,000.) By the 1970s, when Frank was a young veteran moving to the former logging camp, the area had become a run-down summertime resort past its glory days, with good hippie parties and cheap rent in the wintertime.

When I moved here from San Francisco in 2017, the timber baron's former resort was a majority white, working- and middle-class neighborhood of about eight hundred mostly full-time residents. It was also, by all accounts, a fire trap. Houses were wooden, tucked into the edge of forestland on narrow, improvised streets with little or no open space around them. I called it rur-burbia because we had across-the-street neighbors but no sewer service, natural gas lines, or mail delivery. Electric systems were often legacy and sometimes jerry-rigged; woodstoves were used for heat. And of course there were the trees, the reason most of us wanted to live here.

While the redwoods received top billing in the western woods, they were not alone. I had variously heard the ecosystem in which I lived described as coniferous forest, coastal redwood forest, and mixed evergreen forest. The trees reflected that spread. Here, redwoods intermingled with California black oaks—wonderful shade trees in their own right and providers of essential nourishment for many critters. Douglas firs were common in these woods; the coniferous, evergreen trees were almost as tall as redwoods but drier and more avaricious in their growth patterns. Lanky madrone, bay laurel, and tan oak trees occupied the middle ranges of the canopy, and the forest floor was often crowded with huckleberry and hazelnut shrubs. The trees drew to them people who loved the outdoors, people who wanted privacy, and people without the means to live somewhere sunnier. The trees made shade and oxygen, and they ate carbon dioxide. They danced and rustled in wind and rain, and they composed infinite rows of ridgeline after ridgeline decorated in rich greens. They were magnificent.

The trees' futures, like those of people living in the WUI, were precarious. The remaining old-growth coast redwoods were thought to represent 4 percent of the original population. Coast redwoods thrived in moist tracts of forest, often shady gulches and northern slopes, where the trees could drink fog in the summertime. A 2020 study found that climate change could render significant portions of current coast redwood habitat unlivable for the trees as soon as 2030. All over the world, human action and inaction was causing trees to die, burn, become subject to blight and pests, or simply fail to reproduce fast enough to keep up with the rapacity of extraction and settlement. By 2020, the Amazon rainforests—often called the lungs of the planet, although that phrase undersells the region's global importance—were already approaching a tipping point beyond which there was no return, a loss driven by the slash-and-burn operations of illegal mining and logging on Indigenous land. In areas of the Amazon that had already undergone clear-cutting and burning, the forest was now emitting more carbon dioxide than it absorbed. In California, wildfire smoke was negating the recent progress in air quality that carbon emissions cuts had made.

Sonoma County had experienced major wildfires every year since I moved there, but none in the western forest where I lived. Because of its proximity to the coast's marine layer and the moist nature of the redwood habitat, western Sonoma County was historically less fire-prone than the drier eastern flank of the region. In 1903, a wildfire had blown south from the Russian River; it was "one of the fiercest known" in the area, according to the local paper. The paper credited the coastal fog with helping bring it under control within several days, and it stopped just south of the logging camp. In 1961, a large fire burned from the west to within five miles of the former logging camp, but it halted one watershed over. The fire was sparked by equipment belonging to Central and Northern California's utility company, Pacific Gas & Electric (PG&E), which often caused catastrophic wildfires. At the time of the 2020 lightning fires, PG&E had just been convicted of eighty-four counts of manslaughter stemming from deaths in a 2018 fire, but the corporation—despite repeated wrist slaps, fines, and custodianships by various agencies—was mystifyingly still conducting business as usual. Frank had a hazy teenage memory of being sent into the woods with a backpack water sprayer to help fight a fire during the 1960s, but I could never pin down the specific occurrence. Settler fire memory was short; it didn't include Indigenous knowledge of historical events. There were burned-out redwood stumps ten yards from the house in which I lived, but nobody could tell me why.

Frank was right about this: as far as he and I knew, since the time this land had been stolen for us, our neighborhood hadn't burned. That ashy Tuesday evening, between online bouts of fire news, I came across a map overlay of recent fire history in Sonoma County. Almost every area in the county was shaded in a different pastel color, corresponding to the year of its last burn, most of them from the past decade. The only empty area was the one where I lived, vulnerable in its unshaded background. To me, the contours of the map made my forest look like a target, not a shield.

On Wednesday the evacuation warnings began, a steady stream of beeps and buzzes that I followed through a variety of channels, online and on my phone. Friends and relatives began to reach out: my mom and brother, who both lived in Southern California; old friends from New York and San Francisco. The evacuations spread. Our designated zone wasn't yet *mandatory*—get out now—but by Wednesday afternoon we became *warning*—be ready, like, seriously. Small towns to the north and east of us were already *mandatory* and we'd be next, if the wind held. Max was inside our spare bedroom on a conference call. I texted him from the living room: Get off the phone. Time to go.

After I had watched too many red-tinted videos of evacuation gridlock in other wildfires, and after the mess of the countywide Kincade Fire evacuation the previous year, Max and I had crafted our evacuation policy: if given the opportunity, we would leave early, before *warning* became *mandatory* and before traffic became deadly. After the pandemic barred us from friends' homes, we updated the plan: in case of an evacuation, we would camp in my dad and stepmom's backyard in the beach town of Santa Cruz, a three-and-a-half-hour drive south. Santa Cruz was the safest place on Earth I could imagine: my sweet hometown, where the forests kissed the beaches on foggy mornings and their tender whispers lulled you to sleep at night.

As I packed more stuff I realized the fault in our plan, which was smoke, which was everywhere. Tuesday night it had spread across the region like a weighted blanket, seeping into our garden and through the house's old cottage windows. I had asthma, and every time I inhaled, along with irritated eyes, a sore throat, and a headache, a sense of claustrophobia began to set in.

I was wrestling the grown-up folder, where we kept important papers, into the car when I received a text message from my stepmom, Stephanie, who hated smartphones and rarely used them: We're packing up, she said.

I'd seen the CZU Lightning Complex on the satellite fire maps. It was in the Santa Cruz Mountains, just outside the city. Fire in those hills was a semi-regular occurrence; there were densely populated villages up there, but I had assumed they'd get it under control like they usually did. As Max wrapped up his work calls—Sorry, I have to jump, uh, we're evacuating—I zoomed into Santa Cruz on the heat map. The fire had been streaming rapidly west from the mountains toward the beach. Now it was also flowing south through the mountain villages and directly toward the city of Santa Cruz, where the first areas it hit would be the university, where my dad worked, then the neighborhood where he and Stephanie lived. The fire's outline on the map was shaped like an anatomical heart.

A message with no subject line came in from my dad, whose job at the university was managing an organic gardening apprenticeship program. He wrote emails like a Beat poet, or a boomer. He said:

> university is evacuated
> looking rather grim
> but no details as per where the damn fire actually is
> however they have very limited resources to fight the fire
> i heard they were flying in firefighters from australia
> guy across the st works for santa clara open space and he says
> calfire says -santa cruz is as likely to burn as not
> and they wont be able to fight it in the city
> we left house just as precaution
> hope its still there -we'll check tomorrow

They had slept at a hotel on the east side of Santa Cruz, the part of town farthest away from the fire. Steph texted that they were trying to secure a vacation rental in that area as a place to stay; she invited us to come, too, if we had to evacuate. I considered it. Between my parents and me were two out-of-control wildfires and a long drive over a gnarly, forested mountain pass. Santa Cruz was located on a bay, tucked between the ocean and the mountains. There were only three routes out of town (not counting the water). One of them, the coast road, was already on fire north of town, and the other two led eventually to more fires south and east, in the Salinas Valley and Silicon Valley. It was odd that they were even suggesting this; both my stepmom and my dad were high-risk for Covid, and had been strict about the shelter-in-place. I knew they'd never invite us to break pandemic protocol unless they were desperate, or scared.

I typed a reply to my dad. I wanted to say, we're coming home, but I wasn't sure we would. I also wanted to convey that I knew, I understood, how big a deal this was, what it meant for our family and also for the rest of our lives, our world. My dad loved poetry; I found online the text of a poem by Gary Snyder, a former Beat poet who wrote on nature themes, and pasted it into an email. "Mother Earth: Her Whales." The poem was beautiful. It sang of solidarity among all beings, of "tree people," whale people, human people, in the face of capitalistic colonialism. In it, Snyder referred to North America as Turtle Island, a term that was used in many Indigenous traditions. (If I'd googled harder in my pre-evacuation frenzy, I might have been reminded of the fact that Snyder, a white man, had been criticized, most notably by author Leslie Marmon Silko, for appropriating Indigenous and Asian philosophies in his work.) I retitled the subject line of the email "love you guys" and hit Send, uncertain if the missive was intended to reassure my parents or myself. It was time to go.

The car was packed. I made a quick video of everything in our house, in case we needed it for insurance. My narration was shaky and high-pitched, and at times I sounded like an auctioneer: *This jacket cost $100, the shoes were like $80, there are the vinyl records, about 1,000 of them I think, some rare LPs in there.* The inventory was impossible to complete. The handmade built-in cupboards that came with the house. Inside them our chipped, worthless dishes.

Max, as the unofficial head of our neighborhood fire-safety group, went to make the rounds of neighbors to let them know we were leaving. Across the street, Richard and Jay were packing; Frank was lying shirtless on a lounge chair on his patio, trying to absorb the orange heat of the smoke-filtered sun. I went outside to say goodbye to the flowers and the trees.

I stood in the garden and waited for Max to return. He was taking forever. Above me a black oak wiggled in the ashy breeze. The tree was higher than the house. Its trunk was weighted with experience. Hand-shaped leaves adorned tendril-like branches and their silhouettes created non-repeating patterns against the occluded sky. Since I moved to Sonoma County I had been trying to learn to better read the trees. In his 1949 diary *A Sand County Almanac,* ecologist Aldo Leopold chronicled the life span of an oak tree that was felled by lightning on his Wisconsin farm. As he chopped it into firewood, Leopold interpreted the history of the land through the tree's rings: drought years, fire years, eras of human interference or low rabbit populations. When he burned an eighty-year-old tree, the heat it released would be the heat of "eighty years of June suns," he wrote. It wasn't a metaphor. The process of photosynthesis in effect stored the energy of the sun inside a tree's wood; when wood met heat and a fire was sparked, that energy was released. During its life span Leopold's oak had harvested eighty summers of sunlight; by burning it, Leopold gleaned the sun's heat a second time. I eyed the oak above me, and I wished that instead of reading a tree's past, like Leopold, I could see the future in the corpus of this place, could know whether tonight would be the one when the heat of thousands of summer suns was unleashed.

What I most feared to lose in a fire, what the insurance company would never want a video of, was my garden. The lot was about half an acre, half of which was relatively flat. It was open to the sun but encircled by redwoods. The flat area was a lawn, but we'd been sowing wildflower seeds each fall and encouraging it to return to meadow. The other half of the yard was a steep but usable hillside overgrown with non-native blackberry, English ivy vines, and wild-seeded oak, maple, and plum trees. The flower garden was at the crest of that hill; a wooden arch draped in jasmine and climbing roses served as a portal between the beds and the less-groomed hillside. Just below the arch, in roughly defined tiers on either side of a path that Max had made from wood chips, grew six juvenile apple trees, two plum trees, and an Asian pear. The oaks bent down to them. The monolithic redwoods and the dwarf fruit trees lent an odd sense of scale to the garden. Little tree, big tree, miracle planet.

Each year we hewed new terraces and beds out of the wilder part of the hill, pushing back the insatiable vines but keeping natives and volunteers. On the slope sword ferns grew freely. Native raspberry and thimbleberry bushes vied for sunlight among thin tan oak and bay sprouts, which we would manage until they became bush-sized, then prune back. In early spring the entire hill erupted in sprigs of blue and pink: volunteer forget-me-nots and bleeding hearts. Scrub jays lived in the wild plum trees, Anna's hummingbirds patrolled the lower branches of the redwoods, and a collective of acorn woodpeckers occupied the canopy.

From the backyard I could hear Max's voice as he talked with a neighbor. I walked slowly through the flower beds. With Max's labor and my vision this porous edge of the forest had somehow become a garden. Ash fell on blooming dahlias. The marigolds were just coming into their golden orbs. A bed of pink cosmos and motley zinnias was drying out; we were letting them go to seed. The rosebushes looked thirsty. I visited the apple trees and touched a leaf on each one. They had yet to bear fruit.

Before we signed the papers that declared this part of the land our property, in 2017, I had rocked in a hammock strung between two of these redwoods and tried to decide whether this place felt like home. I realized then that the yard reminded me of my dad's teaching garden at the university, where as a kid I often played in summertime while my parents both worked: a steep hillside with ample sunlight but surrounded by a forest of giants. Over the university garden's fence there was a small grove of redwoods, which had been at one time sprouts of a larger central trunk, felled long ago. I loved to lie in the middle of that fairy ring on my back in the itchy duff, gazing up at the sky. I spent a lot of time alone as a child; memories from then were often quiet ones. The cool enclosure of those trees, and the sky viewed from within their cradle, always felt calmer to me than the rest of the world that existed outside my own head and body. Inside the fairy ring I felt sheltered but not constricted. My body was free from the constant obligations of school and home, and my mind could wander as far as the sky could stretch. I felt easily, seamlessly, at home. All this was what I feared to lose in a fire, what I mistakenly thought I possessed.

A MEMORY. It's my first winter in the house. I'm standing with Max in our yard and surveying the apple trees. Deep in dormancy, the young trees look like sticks jammed into the rocky soil. A few small branches—really just twigs—extend uncertainly from their trunks, which themselves are the diameter of my thumb. Although I have never pruned an apple tree before, I am writing a book about how to grow fruit trees. The book is a collaboration with my dad, who in the organic horticultural world is well-known for his teachings. Although I grew up around my dad's garden, I had lived in big cities since the age of sixteen and had not inherited his deep stores of knowledge about plants and how to grow them. Max and I have planted six bare-root apple trees, struggling to get them in the ground between winter rains. My intention is to use this nascent orchard as a way to fact-check the manuscript—a garden as a proof of concept. Above us the redwoods are heavy with moisture. Below them, big-leaf maples extend bare limbs, and beards of lichen dangle from the oldest branches. The ground is wet, too, and the knee-high bell beans and vetch cover crops we sowed in late fall are laced with spiderwebs. It is deep winter, and it is time to prune.

In our book, my dad writes that although people like to think of gardening as an act of creation, at its core all horticulture is a form of intervention. Pruning is one key way that people intervene in fruit trees' reproductive cycles. For a plant, reproduction is survival. Left to its own devices, an apple tree will reproduce wildly and in any way possible. It does this primarily by growing copious amounts of small, not-delicious fruit, shuttling its resources into the production of seeds rather than flesh or flavor. Those seeds will produce trees that are related to, but genetically distinct from, the parent tree. But humans, with our appetites and aesthetics, don't actually want that. We want big, sweet fruit. We want consistency. A Fuji should act and taste like a Fuji, every time. And so we manipulate the tree's division of resources in part by pruning, once in winter and once in summer. Even for those accustomed to it, the act of winter-pruning a deciduous tree can feel contradictory. The basic concept is: the more of a branch you cut off, the more the tree grows back. Cut to grow. Cut to shape. Cut to renew. I understand these principles of fruit tree cultivation, but it still feels wrong to me, to take so much wood from such a small tree.

For me it has been a long season of not trusting human interventions into biology. The winter of 2018 is cold, for California, and I am wearing a hooded sweatshirt, a down vest, and a wool beanie. Beneath my sweatshirt, the waistband of a pair of leggings designed to look like jeans is folded down below my belly, which is swollen. Velcroed around my abdomen is a white elastic brace stained with my sweat and blood. Beneath that, my pelvis is patchworked with incisions and bruises. Beneath those: pain. I have been cut, too, and I have yet to renew.

AS I STOOD in the garden that fiery August afternoon remembering all the other times I'd stood there, the smoke grew heavier. I unwound a hose and began to fill the birdbaths. I had heard anecdotes about wild animals seeking water in the yards of burned-out houses, after their own habitats had burned too. I could have been using this time to better prepare the house for the possibility of fire, running a hose on the roof or sweeping dried leaves off the deck, but it felt urgent to try to lessen the potential hardship of the other critters who lived in these woods. This land had given me so much. It had witnessed what happened to me.

In the months before the lightning storm, I had been trying to write an essay about what happened to my body beginning when we came to live here, and how this land—my garden and the forest around it—had become my companion in damage and renewal. My drafts came up short. I didn't know how to make of my experience a straightforward narrative. I was bored with the metaphors available to me, and the grisly details of how I had come to be cut open seemed certain to overwhelm any casual reader. On some level I must have wanted to protect others from my story's heft, to manage their emotions. With friends and colleagues, too, I struggled to convey what had happened. When I was feeling protective of my own emotions, I would say simply that I had a hysterectomy and the garden had helped me to heal. But I didn't only have a hysterectomy. And I wasn't exactly healed.

What happened was that in May 2017, the same month we began the process of moving from the city to the woods, I had an alarming episode of severe pain during my menstrual cycle that was followed by pelvic muscle spasms, swollen breasts, and bouts of nausea. I thought I might be pregnant. At the time I used birth control in the form of an IUD, a small plastic T-shaped device wrapped in copper wire that sat inside my uterus and, studies showed, safely and discreetly prevented me from falling pregnant 99.9 percent of the time. Still, there was that less-than-1 percent. I visited a gynecologist, who confirmed I was not pregnant and sent me for a transvaginal ultrasound, which showed that my IUD had shifted into the lower aspect of my uterus, where it was neither effective nor safe; however, she said, it was easy to remove, a routine office procedure. When she tugged the IUD out I felt a sudden, intense flare of pain and my doctor made an exclamation that sounded like, huh!? I heaved myself up to a seated position beneath my paper blanket. She disclosed to me the startling fact that only part of the IUD had come out of my body; the little plastic T had, apparently, broken, and one of its arms was still somewhere inside me. This never happens, she told me, it's so weird.

Other things that never happened: the IUD arm was found to be fully embedded inside the wall of my uterus. I underwent a procedure called a *hysteroscopy with or without D&C,* before which I asked for pain medication and was told to take four Advils and nothing stronger and during which my cervix was dilated and a camera/probe inserted through it into my uterus, in an attempt to locate the rogue IUD piece. Over the course of an hour, this process was repeated three times in a row without success. Intermediately, an implement shaped like a butter knife was used to scrape out the lining of my uterus—this was the C part of D&C, also called curettage, a procedure commonly administered under anesthesia in cases of miscarriage or abortion. By this time the flimsy barrier of the Advils had disintegrated and I began to experience more things that had never happened to me: I blazed in and out of feral vocalizations, discovering the existence of previously untapped pain receptors inside my organs, while the floor of the exam room filled with used plastic heat packs and the IUD piece remained landlocked within my myometrium. I was then sent home in shock to bleed and shake while my doctors investigated other surgical options. The IUD was dangerously close to a major blood vessel, which made surgical removal tricky, they said. The next day I began to experience intense stabbing pains inside my pelvis, combined with a feeling of unrelenting pressure from within.

During the following several weeks the pain grew in intensity until I could not sit or walk. I visited eight different healthcare practitioners, none of whom proffered a diagnosis, until I was unable to move my body at all without agony. Eventually, a difficult-to-schedule MRI showed that I had an infection called a perianal abscess, an unusual development that I was told carefully by a new gynecologist was *separate from, but not unrelated to* the rough hysteroscopy procedure, and after thirty-two days of fomenting inside my body the abscess burst. It was an atypical sequence of events, the doctor said. Abscesses rarely happened in this way.

What happened more frequently was that when an abscess burst it created a small tunnel of infection, a passage called a fistula, which in my case began on my perineum, passed through both my internal sphincter muscles, and ended inside my rectum—and which necessitated the attention of another specialist. The new doctor performed a surgery under general anesthesia in which a length of surgical thread called a seton, a treatment invented in medieval times, was looped and knotted in a tight circle through the infectious tract, preventing the fistula from closing before it healed and causing astonishingly painful muscle spasms every time I took a step. After the infection had improved but not healed, a second surgery replaced the seton with a second device, a *cutting* seton, which was made of a sharper type of material and which I was instructed to pull on several times a day until—like a wire cheese cutter, said my surgeon, grasping for analogies—it sliced its way out of the deepest and most tender parts of my body, laid open the infection, and came free. This took four months.

The broken IUD, however, was still locked inside my flesh. As soon as I had recovered enough to be declared fit for another round of anesthesia, my uterus, fallopian tubes, and cervix were removed from my body along with the object inside them. Two weeks later, I hobbled into my backyard and pruned my first apple trees. Within a season my scars went away but my pain never did. And so I suppose it was understandable that I no longer placed faith in expected outcomes. In my experience, every intervention had failed; everything that never happened, everything that everyone had told me would not go wrong, had done so.

Whenever I felt angry or fired up about my situation, I might tell this story differently: As a person capable of becoming pregnant, in order to exercise bodily autonomy, I had to choose between several products, each of which had potentially harmful side effects. And because I lived in a time and place where medicine was a for-profit industry, the products had a twofold purpose: to prevent women from becoming pregnant, and also to make money. Sometimes they hurt women too. Either way, the money was made. Because of the structure of this for-profit industry, the care professionals who tried to heal me were rarely equipped with the resources, research, or incentive necessary to doing so. After my injury became obvious, I was shuttled between specialists who were encouraged by their employer to focus only on a certain several inches of my body, without looking at the whole picture, which incidentally led to further injury. I became collateral damage. Ultimately, because I was neither pregnant nor seeking to become so, the parts of me in which a pregnancy might occur were considered disposable. During the same time period that I was undergoing the side effects of my chosen product, the brand was in the process of being sold by one pharmaceutical company to another for $1.1 billion. The following year, the new owner's reported revenue rose 18 percent. The revamped marketing materials for the device touted its universality, stating that it was *The Only One for Almost Everyone*. For me, it was also the last one.

In subsequent seasons, as my body continued to react to what had happened, I became intensely invested in tending the garden. And I worried intensely about hurting the plants. The process of intervention was always complicated; it wasn't a simple good/bad dichotomy, although some acts were obviously hurtful. Just as the forest around me was testament to the harmful interventions of settlers and the less harmful interventions of Indigenous people before them, my well-tended garden was a dynamic specimen of long-standing cycles of care and injury. The slow, repeated act of being in direct physical relationship to a piece of land was beginning—slowly, repeatedly—to teach me to understand the physical world in terms beyond *well* or *unwell, fertile* or *sterile, whole* or *broken*. While I learned to intervene on the community of soils and species I called a garden, my body tried to heal from interventions by the conglomerate of science and profiteering called healthcare. These dual experiences, which were to me inextricable from one another and from the land on which they occurred, decimated the assumptions about the natural world that I had nursed, unknowing, until someone applied a scalpel and everything hurt, and I realized I was an animal, too.

ANOTHER MEMORY. That first winter of pain, the first year of this garden. The pruning is done. The cuts on my skin are scarring. Max moves a lawn chair into the center of the yard, in the one spot of winter sunlight that still makes it over the redwoods at this time of year. I sit in it when I feel well enough to sit, head tilted, and watch the light move as time passes. I scrutinize the apple trees, looking for signs of the break from dormancy. For a long time, I observe nothing happening. I see the forest cold with dew before the low sun arrives from the east. I see the large white spider that lives in the volunteer rosebush. I witness the stillness of the trees, sticklike apples and girthsome redwoods alike; the oxalis below, their new lavender blooms still closed and turned away from the light. For weeks and then months the apple trees show no sign of breaking the torpor of winter's stillness. Until one morning in late April, the Belle de Boskoop—an heirloom Dutch variety that I chose on my dad's recommendation that it would make good cider—stirs. Halfway up its primary leader branch, on a one-inch stub I'd reluctantly headed back to a single vegetative bud, a hint of a new shoot juts out above the bud. The green curlicue is tiny but luminous; the leaf seems to me to offer the entirety of renewal in its veins. Its particular hue brings to mind the phrase with which the artist Frida Kahlo once defined the color green: "good warm light." The apple tree renders me joyous, for the first time since everything happened. Here it is: light and heat, goodness and growth, a reassurance that I sorely need but haven't trusted; proof that you can cut something and it can still be alive.

SO WHEN I THOUGHT about the possibility of the little wooden house on the hill burning down, I wasn't thinking about my stereo or my family photographs. I was thinking about that good warm light, the relationship I had built with this half acre of land, the feeling I had that was almost inexpressible but might be approximated by saying that this place, those trees, that garden, had helped to heal me to whatever extent I could be considered healed. Decades after the afternoons playing in my dad's garden I had once again become enmeshed in a direct, sustained relationship with what some people called the natural world but was in fact everything.

Max found me in the garden, talking to the trees. We coiled the hose and didn't look back. Our white hatchback carried the go bag, some plastic jugs of water, my mother's guitar from the 1960s, my grandfather's typewriter from the 1940s, and a box of letters written to Max's great-grandfather from his family back in Ukraine after he immigrated to Chicago in the 1910s. We backed out the gravel drive, bumped down the narrow, dead-end street, and bottomed out in the usual place where tree roots raised the asphalt into a burl. Down to the steeper, single-lane road and down down down until we hit the winding, two-lane highway. Two bay trees kissed overhead, forming a tunnel over the car that I envisioned erupting in flames as Max drove calmly. I put a Peter Tosh CD on the stereo, to chill us out. Tosh sang about waters from heaven as we checkerboarded west through dried-out cow pastures. We were early; there were few other cars. Plan enacted. When we reached the Pacific Coast Highway, Max steered south toward the thick sky, and the combusted remnants of the natural world fell on us from above like an early, phantom rain.

· 2 ·

Sea

The road to Santa Cruz wended up from oaky hills and cities shaped like suburbs. Two lanes in each direction coiled over a series of switchbacks through the mountains that separated the city of San Jose from Santa Cruz, a university town of about sixty thousand people on the north side of the Monterey Bay. The pass was dangerous even in fair conditions. As San Jose and the rest of Silicon Valley had grown during the course of my lifetime, the road had become over-trafficked. Today it felt abandoned. Most of the vehicles we passed were pickup trucks, with their innate air of utility. Max drove slowly, but every curve railed the passageways of my nervous system. At each turn I bent my neck to look as far as I could out the window without opening it, striving to see what chaos we twisted toward. The scenery, so familiar from a lifetime of trips back to my hometown, was almost unrecognizable. Everything was shrouded in smoke.

I knew that to my right beyond the immediate tree line were mountains topped with the jagged silhouettes of conifer trees. The ridges chewed like saw blades at the sky's opaque expanse. The Santa Cruz Mountains, part of the Pacific Coast Ranges, ran from the San Francisco Peninsula in the north to the Salinas Valley in the south. Because of the way the coastline was shaped, the mountains bordered Santa Cruz on two sides: east, from where I was approaching; and northwest, where mountains met ocean along the Pacific Coast Highway. The range continued south of town too, swooping west and narrowing around the farmlands of the Pajaro Valley. These hills were home to hundreds of miles of diverse, mostly forested habitats, some of them inaccessible to humans. Where settlers had made inroads there were smatterings of mountain towns, which were nominally part of Santa Cruz but distinctly rural in character. These former logging, railroad, and recreation areas were now home to a mix of wealthy and working-class communities, including the towns of Boulder Creek, Ben Lomond, Felton, and my birthplace of Bonny Doon. West of the mountains, dipping down to the Pacific like a great green bowl, lay the eighteen-thousand-acre forest of Big Basin, the oldest redwood preserve in the nation and the first state park in California.

I couldn't see any of it. My forehead smudged the window. Somewhere, on one of those crests, beneath those big trees that grew to the edge of the continent, I had been born. Somewhere in those woods, my brother and I had played in dirt and run dusty through underbrush, the scent of eucalyptus and the heat of poison oak lingering in our bodies. Somewhere, over there, the shadows of redwood giants had patterned across me as I sat in the wayback of my dad's car on weekend drives, and somewhere in there my dad and mom had loved each other at some point. Beneath trees now obscured by smoke there were wooden houses that stayed chilly even in the heat of summer days. I held trace memories: grown-ups, younger then than I was now, around rustic kitchen tables. Piles of kids napping in bedrooms late at night. Raccoon and deer families rustling outside. I knew, although I had yet to understand, that some of this, maybe all of this, nobody knew how much, was now on fire.

From the car stereo Bud Powell's piano seeped cool reassurance; we'd switched from reggae to jazz. Just before the summit we passed a stretch of cleared forest that cradled a reservoir, its waterlines low like they all were lately, and the smoke cleared a bit. Max asked if I could see anything and I said no, but then I did: several ridges west, between the mass of the mountains and the mud of the sky, I saw flames. Or at least I thought I saw them. There had been an umber color, a non-organic flickering quality to the sky, but I couldn't be sure; the car was in motion, and my sense of scale was skewed by smoke and stress. The flames, if they were flames, seemed to me to be about as tall as the trees on the horizon, which meant they were giant. I thought I saw them, I may have seen them, and on some level I also willed myself to see them. If I witnessed actual fire, it might be evidence that what was happening around me was real, even as the events that were unfolding refuted any logic or structure that my conscious mind could process. If I saw flames, it would mean I was inside a disaster instead of perpetually dodging or anticipating one, and being inside a disaster meant action, meant steps, doing instead of waiting. I was good in emergencies, solid under pressure. If I were inside this disaster, running toward it and not away, maybe there would be something that I could do. Maybe I would feel better, somehow, about all those trees that I couldn't see out the window, dying as I didn't watch.

I could taste them. The trees and all the things made of trees, the structures that underpinned the built world, alchemized in the feculent brew of this smoke. It wasn't only the aftertaste of incinerated wood that I smelled; it was propane tanks and fiberglass insulation and bleach and plastic and steel. Houses, cars, beds. The bodies of bobcats and cougars. All these we breathed. They surged through the closed outer vents of the car and bloomed as an unsteady pain in my forehead, a red fogginess in my eyes, tightness in my chest, that perpetual, occasional cough in the back of Max's throat that he never had before.

It had been four days since we evacuated our home, days of waiting, the scope of the fires infinite yet always just out of my line of sight. The first two nights we had stayed about an hour south of our house at the home of my childhood best friend, Rhys, where she lived with her husband and their two teenaged kids. Technically California was still under a stay-at-home order, to prevent the spread of the novel coronavirus. In the previous five months I hadn't set foot in any house but my own, but it felt disconcertingly normal to walk inside Rhys's family room and pile on the couch to watch TV. We were breaking Covid protocol, betting our mutual health on the nebulous awareness that Rhys and her family had barely left the house since the pandemic began, and Max and I lived in a place where there were few humans at all. We pretended this was safe, but we knew it wasn't. Danger and risk were relative, I was learning again and again that year.

In Marin the heat and smoke had trapped us inside and kept the windows closed. Everyone was working remotely. Every morning four adults and two teens retreated to different corners of the house to partake in their respective video calls for school and work. Rhys had just finished a doctorate in psychology and was in the final stages of becoming a licensed therapist; her husband worked in tech. In the afternoons, Max paced the concrete patios of Rhys's tiered backyard like a recurring blip out the back window, talking on the phone. His job organizing with low-wage service workers had changed since the onset of Covid. Instead of direct organizing with workers, he now worked mostly on video meetings, which, as he put it, sucked. My job was already mostly remote, so for much of the first day of my 2020 wildfire evacuation I sat on the living room couch with Penny, Rhys's mop of a dog. Penny and I tracked fires online. Dozens were burning across the state, none of them contained. I noticed a stack of Rhys's psychology books on the couch next to Penny. At the top was a textbook titled *Reclaiming Your Life from a Traumatic Experience*. I petted the dog, avoided the book.

Rhys finished a client session and came to join me. I said something like, Dude, Bonny Doon is on fire, and she said something like, Oh, yeah, the mountains are already toast. We had been friends since I was six and she was seven and we bonded in ballet class after school. We had always been different from each other: she was lithe and dark and staid, I was round and pink and bohemian. She was a mom, and I was not. We occupied different tax brackets. But we were loyal, and we always told each other the truth. Our friendship was anchored by our hometown, a place with an outsized persona and a funny way of dominating the identities of people who grew up there. The idea of that place burning was too big to encounter, so we set it aside. We moved on. We talked about the dog. The fascist president. Our mothers.

I asked if it was crazy for me to go to Santa Cruz to see my parents, considering the fire might enter the town. We don't use that word anymore, Rhys said wryly, referring to "crazy."

Steph actually texted me, I said. On an actual cellular telephone.

It sounds like your parents really want to see you, she said, already too good at her job.

That first night of evacuation, the winds blew. In Santa Cruz they blew the CZU Fire into residential mountain neighborhoods and ancient redwood groves. In Sonoma County they whipped the Walbridge Fire farther south toward the river, and the power in my neighborhood went out. In my garden, ash flurried over and into every surface, organic or man-made. Redwood duff wafted in the air like embers. Down the hill from the house in which I lived, around midnight, a bay tree cracked and hit a power transmission line and ignited, blazing into the dark sky. Maybe the VFD arrived in time or maybe the line wasn't live anymore, I never got the full story from the neighbors who lived closest to the ignition spot and had seen it, but it didn't spread. Those neighbors had moved away by winter.

On Rhys's foldout couch, the same wind whisked unfamiliar creaks of a strange house to me as I tried to sleep beside an air filter that was working overtime. I got up and cracked the window despite the smoke. The atmosphere was pregnant with doom. Max slept. My lungs tightened around inflamed airways. Electric pain flickered inside my left buttock. A sense of wild claustrophobia took hold of me. I felt as though in the room with me was the weight of all the wood, a million summer suns across millions of acres, releasing its heat into the unpredictable wind. I wanted to run outside toward the looming sky, to be always capable of free motion, and I also wanted to never again go back out there.

In this way passed the first few days of my evacuation. The heat convected under the smoke. The walls of the house felt both thin and restrictive. Everybody did screens. I looked at old pictures of my garden on my phone. Time collapsed, the way it had a habit of doing lately, and all our couch sessions and meals and worries became indistinguishable. Smoke, work, fire maps, work. The world had never been like this before; we had always lived this way.

Evacuating from a disaster was never simply a matter of removing oneself and staying put until it was safe to go home again. Evacuation was a constant state of motion, an evolving equation of risk, comfort, and resources. If people, like Max and me, had friends with houses outside the evacuation zone, they stayed with them. Rich folks and poor folks alike decamped to hotels in their respective price brackets. Covid had made the calculations more difficult. Few shelters were open this year; emergency workers were handing out hotel vouchers, but most hotels were full or closed. Even for us lucky ones, the ones who left early and had somewhere to go, maybe you didn't want to burden your friends, or maybe it was just too crowded, or I couldn't handle one more night in the TV room with the air filters, and so you moved on. Next friend, next landing spot, for however long. Should we go to Santa Cruz, more friends' houses, another friend's barn, back to our own bed? How could we say any of these places were the right ones to run toward?

The wind was fickle. Max texted with a neighbor who hadn't evacuated and learned that the sky over the former logging camp had been smoke-free for two days. Evacuation orders were expanding but our street remained *warning*, not *mandatory*. I wondered if we had overreacted by leaving. There was no way to know. People didn't evacuate and died sometimes; in every major fire there were stories of residents being awoken by walls of flames outside the front door. More frequently, people left and the fire never came. But wasn't that the point of evacuating, for the fire not to come? As I weighed bountiful options, other families were sitting on cots in motels and fairgrounds, waiting to hear whether they still had homes.

Max and I left our carful of possessions in Rhys's garage and took clothes and the go bag, all we'd need to spend a night somewhere, we weren't sure where. Max turned north at the highway almost out of habit; to home then, just for a minute. We could make sure it was still there. Water the garden. Then if the fire was still spreading, we'd go to Santa Cruz.

The little white house and big red trees were still there. The electricity was back on. Frank was in his yard, sunbathing again. The sky was blue, that lying blue I loved so much, a blue that said, everything may be wrong in the world but right here, on this patch of hill, we're okay. The blue invited us to spend the night, and we succumbed. It felt safe at the house, although it wasn't. In the morning from the garden I could look north and see the smoke cloud rising like a strange planet over the tree line: our new neighbor, the fire.

Santa Cruz, then. Family. We left the range of our fire and drove through the smoke of other people's fires—LNU, SCU, CZU, so many letters representing so many catastrophes—and now I was getting carsick leaning against the passenger-side window and trying to spot flames in the mountains where I was born. The ridgetops sawed into the impenetrable sky and stalled. Inside my body, my pain was deep and radiant, a diamond unfound. Through every crack in the car's body the smoke snuck and thudded into me, red-tinged and sour-tasting. The smoke made clear: we were all inside the same disaster.

Any origin story is at heart a romance, a love story about oneself. The origin story of my California, the forests I thought of as mine, and the fires that occurred there, began in Santa Cruz. I was born in the mid-1970s, in a trailer in a redwood forest. My dad delivered me because I arrived before the midwife. The trailer in Bonny Doon was next to a half-built geodesic dome on twenty-two acres of land; we were only living in the trailer until the dome was finished, which it never was. I had no memory of the place—we left when I was a baby—but when I visited decades later the dome was still there, unfinished, torn bits of plastic tarp fluttering from its disintegrating wooden beams.

At the time my parents lived at the dome they were in their mid-twenties. A photo of my father from these years showed him in black and white: unimaginably youthful, he stood waist-deep in brown grass and held an acoustic guitar. He didn't look at the camera; his sandy-blond hair covered his line of sight as he focused on the instrument strapped to his chest. In the same era of photos, my mother had long straight black hair and always appeared to be in a sheen of sunshine, a sort of faraway focus to her malachite eyes.

Instead of paying rent at the dome, my father milked the landlord's small herd of goats twice a day. To make money, my mother grew flowers in a clearing in the mountain sun, arranged the clippings into bouquets, and delivered them for a fee to local businesses, for whom she also did bookkeeping. It always sounded idyllic to me when either of my parents told it, but in reality, my dad hated goat milk and my mom was always worried about money. My parents were not poor—their families had been wealthy (mom) and lower middle class/military (dad)—but like many people in their twenties, they were broke. Life in the United States was cheaper in those years, but still my mom recalled living off sandwiches made from peanut butter and tortillas when she was pregnant with my older brother.

My parents weren't the Haight Street kind of hippies; they didn't tongue LSD or dance in hedonistic gatherings to far-out rock bands. They were the kind of hippies who lived in the woods without electricity and had their babies at home. They had come to Santa Cruz separately in the late 1960s, both seeking a dose of the consciousness-expanding energy that preoccupied their generation. My mom had been a civil rights activist at her preppy private college before dropping out and transferring to UC Santa Cruz; my dad, who never finished college, escaped conscription to Vietnam through a combination of trickery and luck.

Since being colonized by the Spanish in the 1700s, Santa Cruz had been a logging town, a tannery town, a fishing town, and a beach resort. In the late 1960s, a new state university with progressive concepts of education was in the process of transforming it into a radical college town. At that moment in the twentieth century, my parents and their peers—a particular subset of the baby boomer generation—were realizing they didn't have to do what their parents had done. They didn't have to live where it snowed, cut their hair a certain way, go into traditional professions, or fight in a war. Not coincidentally, they were at the cusp of a new cultural moment, in which spirituality and self-regard were thought to lead to enlightenment and peace. Mother Nature, long neglected, was on the rise. The dirty cycle of conformity and consumerism would end with them.

My parents and a small group of friends practiced yoga and followed the teachings of Baba Hari Dass, an Indian monk who since 1952 had lived under a vow of silence. Babaji, as we called him, communicated by writing on a small chalkboard tied around his neck. He had long white hair, warm expressive eyes, and an impeccably timed sense of humor that he conveyed through chalk-scribbled koans and a repertoire of silly noises. Babaji came to Santa Cruz to teach yoga after being invited to the United States by two UC Davis students, who had met him in India, and their professor, who became his sponsor. With the encouragement of their guru, my parents and their friends adopted Sanskrit names. The nascent spiritual community together looked for, and found, a span of three hundred acres in the southern Santa Cruz Mountains outside of Gilroy and used some of their parents' money to buy it, calling the place simply the Land. The Land would eventually house a school, a Hindu temple, and a yoga retreat center. My family didn't live on the Land—the dome was in a different part of the mountains—but they were fully a part of its community. My parents married in a barefoot ceremony in the redwoods; they had children, a boy and a girl, and gave the responsibility of naming us to Babaji.

They found another teacher in Alan Chadwick, an eccentric English former actor and horticulturalist who'd been hired to start a teaching garden at the new university. My dad eventually lucked into a job at the garden, teaching wannabe farmers how to grow apples sweet as plums, roses in the middle of November, purple broccolis thick as carpet in their beds. My mom finished her degree (in Aesthetic Studies, a short-lived cross-discipline major). Her parents offered to help them buy a house, so we moved from the woods into town, and drifted further away from the community on the Land. I had only secondhand memories of this; my parents separated around the same time we moved into town, when I was three. Dad kept the VW bug and Mom bought a used station wagon, and I spent the next thirteen years shuttling between their houses. Around the same time, they changed their names back to the ones their parents had given them. My mom began grad school in theater, immersing herself in the bare literary landscapes of Samuel Beckett and Marguerite Duras. She still did the books for local businesses late at night, the rhythms of her ten-key calculator accompanying me as I learned to read in the big armchair by the fireplace.

In 1980, Ronald Reagan, the conservative former governor of California, was elected president, and the dirty cycles of the world continued unabated. Some of the hippies and dropouts began to tire of being broke. They missed TV; they cut their hair and got divorces; bought condos; tried for tenure. Even in the Edenic enclave of Santa Cruz, the spiritual renaissance of the 1970s dovetailed into the individualism of a new era. We got a television set, although we were only allowed to watch it on special occasions. When I was eight, my dad remarried, to Stephanie, an elementary school teacher, artist, and avid outdoorsperson.

Santa Cruz was a place that had the outdoors embedded in its identity. Its stunning natural surroundings spawned several legendary surfboard and skateboard apparel brands, as well as the cultures that preceded the merch. It was home to progressive environmental activism efforts and world-famous mountain bike, wet suit, and organic food companies. The university's mascot was the banana slug, yellow denizen of redwood habitats. The campus was home to robust environmental studies and ecology programs, including a research group that spearheaded the rehabilitation of the peregrine falcon from near extinction in the 1980s. The UCSC Farm and Garden, where my dad worked, served as a field classroom for an apprenticeship program whose alumni were responsible for groundbreaking sustainable agriculture movements around the world. And Santa Cruz was a destination for beach tourism, that most popular form of outdoor appreciation, which in 2019 contributed $1 billion to the local economy.

On weekends my family members occupied their respective outside spaces—my dad at work, my stepmom hiking or painting in plein air, my mom on her deck chair grading student papers, my brother surfing or skating. Like most kids, I rebelled against some of my family's influences. I loved theater and poetry like my parents did, stayed inside on sunny afternoons, wore black, eschewed spiritualism in all forms. By the time I was eight years old I had claimed the indoors—and the very indoor activity of ballet—as mine.

It was through dance that I first understood the complexities of having an animal body. The first ballet teacher I had was a grumpy old Russian lady with a cane, of whom I was so afraid that I once peed in my tights rather than ask her for permission to take a bathroom break. Ballet tamed the instincts of childhood play into regimen, instilling routine and drilling patience into even the most beautiful of gestures.

At the studio there was a cabal of older girls, bony beasts who ran in a pack but didn't seem to be very fond of one another. Rhys and I stuck together. After class we observed as the older students removed their ribbon-tied toe shoes, rolling pink, footless tights down long legs and unpacking their feet from layers of wadded-up cotton. What emerged was a fairy-tale curse: toes crooked as oak limbs and footpads as gnarled as stumps. When they weren't dancing, the older girls walked oddly—their feet permanently turned out and their gait stiff, as though it hurt to put weight on their shins, almost wobbly, like newborn goats. In the dressing room it smelled like sweat and blood.

On off days from ballet class, Rhys and I made our own dances, conducting rehearsals in her living room soundtracked by pop music cassettes. Choreographing outside the bounds of ballet felt naughty, and free. We emulated women we'd seen in music videos, rotating our hips in circles and turning our toes in like petulant pigeons. We flipped our hair around and spread our fingers into star shapes wide enough to give a Russian ballet mistress a heart attack. We took our dances seriously, rewinding and pausing each song to rehearse until we got it right. Over the years, our dancing bodies began to curve and swoop in new ways. Our perfect ronds de jambe took on more swagger. Our chests shook like almost-done Jell-O. We had little concept of what our bodies were, what they meant, what they might endure or do throughout our lives as human women. But we already sensed a simmering power, a danger in them. Our bodies were strong yet breakable, uncontrollable yet precise, possessed of the potential for both pleasure and pain. Just as we understood how to toe the line between exhibitionism and propriety in and out of dance class, just as we would always thereafter recognize the smell of cotton soaked with blood, even as young as ten or eleven years old we knew somehow, subconsciously, that our bodies were still wild things, capable of savagery and beauty at the exact same time.

As a teenager I couldn't wait to leave Santa Cruz, which I perceived as a hypocritically righteous culture bubble, and I spent much of my subsequent young adulthood leaving different places. When I was sixteen, I moved to Boston for college, then promptly dropped out. At seventeen I moved to San Francisco. There were the months in Paris.

Years in New York. San Francisco again. Six different apartments over two years in Portland, Oregon. When I moved back to San Francisco for the third time in my mid-twenties, I hadn't planned to stay. But even as I moved to different big cities, in and out of California, I retained a sentimentality for Northern California's ecologies and landscapes. I never really felt whole when I was more than fifteen minutes from a body of water; I craved fog. I often felt a deficiency in my lungs that I suspected might correspond to the oxygen output of a redwood tree. Even after decades of city-love, the landscapes of coastal California remained at my root. I somehow understood instinctively that between a redwood forest and the Pacific Ocean was the most natural place for my body to be. I had chosen to become a big-city person, but I would never be a person who didn't believe that people were a part of the ecosystems around them, humans were animals, and the natural world was by definition also the human one.

My return to the trees happened quickly once it happened. In 2014 Max was sent to Sonoma County, a couple of hours north of San Francisco, to organize with service workers at a casino; we sublet our apartment in the city and rented a log cabin near the Russian River, a few miles north of our future home. West Sonoma County was graced with similar weather patterns and ecosystems as Santa Cruz. The redwoods, cool in their armored skin, emitted the odor of tannins and something vague that reminded me of home. After a year in the log cabin we returned to the city, which the tech boom had been busy transforming into a place I no longer liked. I soon realized that every glance I gave every inch of San Francisco had somehow become an angry glare, a tardy expression of a grief that I felt strongly for a vibe I couldn't prove had been destroyed, even though its absence was evident in every empty storefront and gray condo flip. My landlord's perpetual threats to sell our rent-controlled building grew more frequent—I'll just sell it, he'd say, if you need the window fixed, if you want the paint above your bed not to peel, I'll sell. Eventually, two software engineers ended up buying the building and, following a lengthy and complicated timeline of threats and negotiations, paid everyone to leave.

I had never imagined I'd leave the city, and neither Max nor I had ever thought we'd own a home—or, more accurately, we never thought we would pay a monthly sum to a financial institution whose name we barely knew in exchange for permission to occupy a building, some old tree groves, and the weird sinkhole in the garden that we kept pouring soil into. People called this process building equity, which to me seemed like an ironic word for it. There was no family money coming to me or Max; our incomes were modest by San Francisco standards. But they were secure by any other standards: unusually, we both had salaried jobs at the time. So we signed a stack of papers we didn't understand, and the papers told us we owned a little house on a hill in the redwoods on stolen Indigenous land—another vacation wonderland for me to call home.

The beach rental at Rio Del Mar was stucco and had once been pink. Now its exterior walls were corroded with scuff marks from years of sun and salt. We entered the building's garage through a back alley. Inside the house, there was the usual soulless feeling of a home that nobody lived in. Wicker and glass furniture of questionable modernity. Unusual utensil choices. An overabundance of the colors turquoise and tan. But there was a balcony, its sliding glass doors closed and caked in sea salt. Through the grime I could see the ocean, calm despite the malevolent wind. Even just a glance at the water oriented me momentarily, as the ocean always could.

The rental cost my parents $400 a night. It was outside of our family's usual budgetary wheelhouse but doable for my parents in an emergency, for a few nights anyway. In the slim galley kitchen the hurricane of my dad had already hit, and the tile countertops were covered in grocery bags of citrus fruits picked in haste from his front yard. Buckets of apples and potatoes lined the dining room floor. There were headlamps and flashlights everywhere, just in case. He'd brought his pots and pans from home. I piled our snacks on the counter with his. Into the blue second bedroom went the go bag, our clothes bags, and a few totes of random items I'd grabbed in a hurry on our stopover at home: Mostly books. Scrabble. For some reason, a small, vintage metal calendar that flipped over every day to display the date. I quickly realized we had packed for an apocalypse but not a sleepover; we didn't have toothpaste or shampoo. Max ran out to the store and returned with both items and an odd half smile: the clerk had asked if he was an evacuee, and he'd said yes, surprised to find that the moniker applied to him, and gotten 15 percent off. Max loved a discount.

We had agreed on the Covid protocol ahead of time: masks inside, distance when possible, frequent handwashing, wipe down everything in the bathroom after you use it. Stephanie greeted us, and soon my dad arrived back from doing errands. We didn't hug.

My dad, Orin, was in his early seventies. His voice was low in volume but high in energy, and he walked with a bowlegged tilt. His surfer-style bowl cut was turning from blond to silver and had grown wildly in the pandemic, giving him the appearance of a mad scientist. Stephanie, in her sixties, was tall and pragmatic and unflinchingly intelligent. She and Orin had two daughters, who were half sisters to me and my older brother, Niranjan. Katie lived in Oakland, and Caroline, the younger, lived nearby. Caroline came over with curbside takeout; we all sat on the balcony, Covid-wary of one another's personal space, but soon lapsed into our usual family-dinner-table mode of conversation. We talked over each other, excited to not be alone.

Orin had brought from his house a six-pack of beer with a label that read RESILIENCE IPA. A local brewery had released the beer as a fundraiser for victims of another unprecedented wildfire in Northern California the prior year. He must have been saving it for a special occasion. We made dark, evergreen jokes about climate change while we drank. There was a toast to Buster, the farm cat at my dad's work who was named after the San Francisco Giants catcher. Buster had been "rescued" from the evacuated campus by a kind former apprentice who crossed police lines to retrieve him; he was now living the luxurious life of an indoor cat at his liberator's house. I asked my dad about a recently retired coworker of his who lived in Bonny Doon; her house had burned down. Max updated everyone on his organizing efforts, and I asked my dad for the smoke forecast. Like most farmers, Orin was obsessed with weather reports. For most of my life I had traced the dreadful progress of climate change through his weather updates. Although wind drove its movements, smoke wasn't technically weather, he said, and meteorologists were having a hard time predicting it. But, he said, the apple trees at the garden had dropped their fruit early, en masse, as soon as the fires had started. It was probably the heat and smoke.

The news from Caroline, who was a flower farmer, was that a famous chef who set up kitchens at disaster sites had arrived in town; Caroline was helping friends at another farm harvest vegetables to donate. After the pandemic lull it felt almost like a normal workday, she said. Weirdly chill.

At some point we ate dinner and Caroline went home. Warmed from the company but drained from exertion, I went to bed. New unfamiliar bed, same late-night doom-scrolling the fires. It felt incongruous to be in this monochromatic vacation house while just miles away fire was remaking the map of my childhood with its strange orange light. Pain bubbled inside my belly. I took a muscle relaxer and went to sleep.

In the morning Max and I learned that some evacuations were being lifted in west Sonoma, while in the east of the county others were being put in place. Our zone was still *warning;* our fire was still north of the river. In Santa Cruz, the CZU Fire had burned through several of the farther flung mountain neighborhoods. Some people had had no warning; they outlasted the fire by hiding in ponds and creeks. On a dead-end road named Last Chance, up the coast, residents had hiked out through the forest as it burned. One hadn't made it, an older man whose body had just been found. I read out loud from the newspaper article and when I was done my dad said quietly, Oh, I knew him. The man had spent time on the Land in the seventies. After Vietnam he had anger issues, my dad said, and he took a vow of silence and got really into meditation. He was an odd guy, but good.

Other stories arrived through rumor and friends. Informal mutual aid networks for evacuees and first responders were springing up. Mountain lions and wild boars were seen in residential areas, singed and scared. The fire had made it to the road I'd been born on, but not yet to the geodesic dome. The CZU Lightning Complex was smaller than other fire complexes burning elsewhere, but it was exponentially more destructive for its size; it ultimately burned fifteen hundred homes. The fire had started as several separate lightning strikes, which after two days united and quadrupled in size when the wind changed.

By that time, firefighting resources around the state were already so strained that when the CZU unit called for assistance there wasn't any help to send. Equipment and personnel were all elsewhere; Santa Cruz had even sent a strike team up to Sonoma County. Municipal fire departments and the CZU unit were fighting the fire, but it was a bare-bones crew. We heard that people were staying behind evacuation lines to lay hoses around their neighbors' houses, using their own farm tractors to cut firebreaks in the vegetation, and setting up informal patrols for flare-ups at night. In the old-growth redwood preserve at Big Basin, fire had laddered up into the canopy for the first time in living memory. The entire park had burned. All in one night.

After the bad news, the emails: work, scroll, work. In the few days since the fires started, the media had had time to catch up. The discourse was full of reported articles, think pieces, and personal essays about the doomed future of California. *If you like 2020, you're going to love 2050,* one pundit wisecracked. These fires were bigger and more destructive than previous fires. The scale of ecological calamity was insurmountable. Young professionals were talking about leaving. Climate scientists were struggling with suicidal thoughts. The fires were only going to get worse. The fires were going to burn year-round. They would burn forever. They would kill us all. Another article that lodged deep inside my nervous system ended with a Bay Area–based reporter being urged by the climate expert he was interviewing to leave California, just move, *get out now.*

Max took a break and we slipped away for a walk, double masked. The usual haze of summer morning fog had been replaced with an ill pallor of toxicity. The house's front yard was closed off with a cheap wrought-iron gate, through which beachgoers had thrown empty cans and chip bags. Sad hedge shrubs tilted, the soil beneath them dry. Directly in front of the gate was a public parking lot, and adjacent was a corner store. Beyond the parking lot the gray bay stretched thinly in either direction, with the sand a slightly lighter, closer gray. A few hundred yards down the beach a rickety wooden pier, faded from age, jutted into the surf. At the edge of the parking lot a lone palm tree stood like a parody of a California tableau. In the opposite direction was a strip of beach houses that were in the process of slowly falling into the ocean; behind the houses the cliffs were eroding, too. The coast road here had been moved inland since I was a kid. California was like that. The ground was always shifting.

By this point in the climate crisis I'd heard the word *solastalgia* hundreds of times. Glenn Albrecht, an Australian professor and environmental activist, had coined the term in 2004 after observing signs of displacement in a community besieged by coal mining in southeastern Australia. Solastalgia referred to the grief that a person felt when her home environment was irrevocably altered. It was, Albrecht wrote, "a form of homesickness one gets when one is still at home." The idea overlapped with the then new psychological concept of ecological grief, now commonly called climate grief. When I had first heard about solastalgia, I was surprised there wasn't a word for it already because that was how I felt about my hometown all the time.

Long before the fires, like most of California, Santa Cruz had already been a place that understood the power of planetary forces to upheave human lives. In the 1950s and the 1980s there had been devastating floods on the San Lorenzo River, which flowed from Zayante up in the mountains down to the Beach Boardwalk. In winter in the mountains, after rain, after fire, landslides came. In 1982, thirteen people had been killed in a slide in the mountains at Love Creek, then the deadliest debris flow disaster in the state. And in 1989, when I was thirteen years old, a major earthquake had further split the substrata of the Pacific and North American tectonic plates on a point along a small tributary of the San Andreas Fault, not far from where I was walking with Max. About a mile inland from the beach house there was a state park that my mother had frequently taken my brother and me to when we were kids; the park was rarely crowded, and redwoods made it always cool in the summer. Inside this park, up nine miles of trail, there was a small sign posted at a nondescript point. The sign read EPICENTER, LOMA PRIETA EARTHQUAKE, MAGNITUDE 6.9, and listed a date, time, and coordinates.

The earthquake was California's biggest since the 1906 *big one* in San Francisco. It became instantly world-famous due to its live appearance on television, thanks to a concurrent World Series game—between the San Francisco Giants and Oakland A's—being broadcast from Candlestick Park. In Oakland, a double-decker freeway collapsed during rush hour and killed dozens of people. In San Francisco, entire blocks burned in fires sparked by natural gas leaks. But the epicenter was here in Santa Cruz. In the downtown shopping district, one-third of the buildings, most of them made from unreinforced brick, collapsed or were fatally damaged, displacing fifty businesses on one street. Most of that main drag was demolished within weeks. My high school years were spent wandering amid craters in the ground where shops had been torn down; the bookstore where I worked had been located in a tent for several years after the quake.

Within days of the earthquake, signs went up on chain-link fences around the rubble: DOWNTOWN IS OPEN FOR BUSINESS. Because of active and progressive community participation, Santa Cruz fared better during the redevelopment process than many places did after a natural disaster. But the earthquake was an entry point for the process of gentrification. The disaster made way for the type of economic opportunism—real estate speculation, the pushing out of low-income and precarious residents—that many parts of California also saw during the 1990s, but the difference for Santa Cruz was time: it happened here in the span of seventeen seconds. For me, the earthquake brought to light a hypocrisy of the boomer generation. Since the freewheeling days of envisioning a world without commerce, many of my parents' peers had become developers, real estate agents, and business owners. As a result, Santa Cruz was now a more affluent yet less-livable place than during my childhood. Young people like my parents had been could never afford to move there now. KEEP SANTA CRUZ WEIRD was a bumper sticker I saw less and less, lately. In these ways, the earthquake had also been my initial education in disaster capitalism. From that point on in my life, I looked for signs of the machinery of capital at work beneath everything.

Santa Cruz had never been the same since the earthquake. I had never been the same—I spent decades of my life wandering, looking for a new way to feel at home, and only recently had I thought I found it. I was good in emergencies because I had experienced them. I had left my hometown in part because it had left me first.

There were tourists on the beach, although not many, and Max and I steered clear. A few surfers sat in their trucks looking at the waves, which were flat. I noticed the nearby RV campground was full, and I wondered if they were evacuees, but it was still summer and we were in a resort town. I took off my shoes. The corner store did slow but steady business. We walked along the tide line toward the pier, dodging seaweed and foam.

In the shallows at the pier's end, the remains of a huge boat lay sunken in the water. The ship was made of concrete, but everyone erroneously called it the cement ship. It had been designed as a battleship during the wood shortages that plagued the First World War but the war ended before it was used; in 1930 it had been towed here and docked. The cement ship had been at one time a nightclub, a bathhouse, and, in my childhood, an extension of the pier. Locals had walked on and fished from the boat until a few years earlier, when severe winter rainstorms torqued the tides so violently that the ship was overturned and twisted. Now it lay like the remains of a monstrous fish in the water, half-sunk, covered in roosting birds.

Stephanie was the real birder in my family, but I loved to watch them. Squadrons of brown pelicans landed and took off from the boat. I always thought they looked a little goofy when taking wing, because their legs dangled ungracefully in the air. As pelicans looked for fish, skimming the surface of waves while in flight, eager gulls followed them in the water. After growing up on the coast I was so familiar with the sight of seagulls that I rarely paid them attention. They were scavengers, one step up from pigeons. Common. But now that I looked at one closely, it was beautiful. Its feather contours mingled with the mottled hues of sand, surf, sky. One bright speck: a red dot on its beak. Gulls weren't glamorous but they were survivors. Like all seabirds, they were equally at home over land or water, navigating space with a casual expertise that all the boats and airplanes in human invention could never achieve.

Should we go home? I asked Max.

I don't know, he said. Things don't seem very much under control anywhere.

A lone family of tourists played in the shallow surf. My dad, who swam in the ocean most days, had cautioned us not to swim here. This beach had recently become a popular cruising ground for juvenile great white sharks; the teen predators liked to gather in the cement ship's wreckage.

Do you think our house will burn down? I asked for the thousandth time.

Probably not? Max, ever the optimist. He continued: Not this time, at least? Unless shit gets really crazy.

I furrowed my brow at him for jinxing us. With previous wildfires in Sonoma County, Max and I had talked about what it might feel like for our house to burn down. But I hadn't yet contemplated what it might feel like if our house *didn't* burn down, and the fires continued. How would I live with that kind of anxiety? With the smoke? I told him about all the articles.

I guess this is going to keep happening, I said. It wasn't really a question.

Climate chaos. We're in it, Max said. He toed a marooned piece of seaweed, the long ropelike kind I used to use as a jump rope when I was a kid.

It's not even fire season yet, I complained. Should we be thinking of leaving? I mean leaving Sonoma County? Moving away?

The last couple of days, as talking heads had swarmed, predicting an infernal future for California, friends in other places started to ask whether I was going to leave. None of my close friends asked it directly, but in their check-in text messages they were beginning to say things like, So, what do you think you'll do? In reply, I would fill them in on our evacuation plans, ignoring the question beneath the question.

I don't want to talk about leaving, Max said. He was clearly rankled. Apparently, people had been asking him too.

It's going to keep getting worse, I prodded.

You can't run away from climate change.

No, but we could be slightly less on the front lines, I said. I mean, it *is* an option.

I'm just . . . not interested in talking about that, Max repeated. It makes me sad.

I didn't push it further. The concept of leaving willingly the home and land I felt physically entangled with was one more thing my brain couldn't begin to entertain this week. I picked up my own rope of seaweed and began to swing it like a lasso toward the shark boat. I recalled all the previous places I'd loved and left, unsure which of my decisions were the right ones.

What do you think we should do about the trip, Max said.

I had almost forgotten. We were supposed to be going on vacation next week. Max had planned his first-ever solo backpacking excursion, navigating Covid regulations to get a rare backcountry permit at Yosemite National Park. I had plans to hunker down at home in what my friend Lydia, a novelist, called dirtbag mode, which had nothing to do with the similarly named rock-climbing culture and instead mostly consisted of eating junk food, not doing the dishes, and working on my novel.

Do you still want to go, I said. I mean, you could go. Yosemite's not on fire.

I guess so? It seems like it's not too smoky there. And the backcountry is totally Covid safe. No people!

I laughed.

What would you do, though? he asked.

I don't know. I don't think I can be at home alone right now, I said.

It was understood I couldn't go backpacking with him. I could hike, even with my ever-present pelvic pain, but hills hurt the worst. Carrying everything with me at more than ten thousand feet of elevation in potentially poor air quality wasn't an option.

Get a cabin or something, he said. Come with me to the mountains, and just do your own trip. You can drop me off, that way you can come get me if there's a fire or an evacuation while I'm in the backcountry. Just, like, get a cabin somewhere and write.

It wasn't a bad idea. Stephanie could recommend a good spot to stay. She'd spent time in the High Sierra during college, studying botany and leading backpacking trips. We could stop back home first, water the garden, get the rest of the camping stuff.

I just think we should stay on the road a little bit longer, said Max, until it's definitely safe to go back. Maybe pretend we're traveling? Like an adventure.

Adventure was our term for the risky things, large and small, that Max and I did. We had big, limitless ideas about the world and set our sights accordingly—we had traveled in jungles and cities; we had planned political protests and performance events; we had once left our jobs and biked around Europe for six months, living mostly in a tent. We eschewed traditional markers of adulthood like career tracks or children and instead invested in good friends and good food. When there was trouble we usually made it through thanks to a combination of will, skill, luck, and privilege. We were educated, employed, insured, and white. I had a credit card on which I could put five nights at a cabin in the Sierra Nevada and not really worry about it. I was a tough cookie, and Max was unstoppable, a force. We could do adventure.

Max finalized it: Okay, so. It'll be an adventure.

Apocalypse vacation, I said, wry. Cool. Totally normal and cool.

I contemplated the grim sky above the quiet sea. Even with all the resources smoothing my path, I didn't want to be an evacuee or a faux adventurer. I wanted to continue to be an exception to the consequences of climate change. I had long understood that I would always be excepted from certain experiences because I was American, because of my race and class, and because my gender expression aligned with the expectations of others. The world made life easier on me whether

I acknowledged that or not, and regardless of how that acknowledgment affected the way I acted on or used such power. There was another component to my sense of exceptionalism that I hadn't yet fully admitted to myself, but as I aged I began to be aware of: I was smart. I was creative. I noticed things: eagle eye, poet heart. These qualities were important to my identity as a creative person, but they often kept me at a remove. On some level I must have thought my intellect, combined with my privilege, would allow ecological calamity, sociopolitical destabilization, collapse—whatever the cascading disasters of the present era were called—to remain as theoretical future prospects in my life: subjects, not present dangers. This line of thinking didn't hold up. A person could fall victim to disaster whether or not she was a smarty-pants. One had nothing to do with the other. I could try to put a unique spin on it, but my desire to remain an observer of history instead of its victim was banal. It was the same desire everyone had.

There must have been schools of fish out beyond the break, because a flock of brown pelicans came in for a dive-bombing session and the gulls jostled for wave space awaiting their castoffs. Black cormorants with their slick profiles rocked easy on the swells. Calls rustled like wind through the smoke. A feeding frenzy. We ducked under the pier.

A MEMORY. I am a child and when my mom takes us to this beach, I often wander off and play on my own beneath the shade of the pier. My brother surfs; my mom reads books and magazines in a beach chair. I prefer to sit at the tide line where the waves land onshore. I enjoy the sensation of sitting directly on the ground and allowing myself to sink slowly into wet sand until the incoming water almost overtakes me. When a wave approaches I rise jubilantly to stand or jump and greet it, full-face. Salt taste, stinging eyes, crashing sounds. I make a game of it, a war game of sorts—when I play in the surf I always have the distinct impression that I am protecting the land against some sort of invasion. I never articulate just what type of invading hordes I imagine might be attacking—it's not the water, which also needs protecting, but rather some malevolent force borne in on the waves. And I know what I am supposed to defend: the beach, the trees on the cliff, even the restless ground beneath me. My upbringing, hometown, cultural values, and community together reinforce the importance of our protecting this: *Mother earth, her whales.* As it turned out, we didn't do a very good job.

MAX AND I WALKED back to the crappy beach house, leaving our sandy shoes on the doorstep. The gulls followed at a safe distance.

Every morning my father woke before sunrise and made a large amount of coffee, some of which he drank immediately and some of which he reserved in a pint glass on the counter for later reheating. He then selected a yellow legal pad from the stacks of papers and books that littered the kitchen table, grabbed a marking pen, and wrote. He wrote before he read the sports pages; before he loaded up his bicycle panniers with dog-eared horticultural books and those same yellow notepads and rode up the hill to the university; before he walked the paths of the garden alone to survey the progress of the plants; before he witnessed the beneficial insects become active as the fog lifted and the birds descended; before he watered the seedlings in the greenhouse or wrote the day's tasks for the apprentices on an outdoor chalkboard in vigorous all-caps; before he stood on a sloped hillside in running shoes and shorts all day, the way he had for more than forty years, and taught future organic farmers how to grow stuff until the fog came back and the day was over. He always wrote with books at hand. Soil biology, roses, pruning, the history of agriculture, poetry, baseball. What he was producing, I wasn't always sure. A new book idea, maybe; lesson plans for his workshops, which had been canceled by Covid. Sometimes he wrote dispatches for his program's donor newsletter. Mostly I suspected he was just reframing his own thoughts and teachings, over and over again. Revising. When we wrote the fruit tree book together I realized with humorous dread that my creative process was similar: rephrasing and rephrasing myself, then combining, then pruning those phrasings to find their best iteration. It was a laborious process that sometimes felt wasteful, but it usually worked.

When I got up on our third morning in Santa Cruz, my dad, who had likely been up for hours, was already sitting outside on the balcony doing his morning writing routine. I made tea at the tile counter and regarded him through the dirty sliding-glass doors. He wasn't wearing a smoke mask. He sat at the small patio table and stared at his yellow legal pad. A stack of books teetered beside him. To me he looked out of place, out of time, enacting a ritual whose connection to the tangible was quickly dissolving. The university was evacuated and a wildfire was burning toward it. He hadn't had apprentices to teach since the start of the pandemic. He, at age seventy, had been tending the three-acre garden at the university largely by himself, doing what he could to preserve the perennials and protect the plants from the hordes of coyotes, wild turkeys, and deer encroaching on his life's work. As I watched him write, a staggering sadness overcame me. My body felt it first and sent coursing through itself a current of pain. My dad's body was covered in marks of his occupation, small scars and sun spots. His skin looked tissue thin. This was a person of routine, a person who hadn't left California in more than forty years because he didn't think anywhere else could possibly be as good. What must it have been like for him, to witness his environment so painfully altered? Displaced from his house and garden, even temporarily, he looked at sea. The pain rose and pooled behind my lungs.

Since my hysterectomy I had become more attuned to what I felt was an atmosphere of slow decline around me. From politics to ecology to my own physiology, systems were failing in ways large and small. There was no backup plan. No one was going to save us. When I got this feeling, I experienced in my body a strong sensation not unlike that of slipping and falling. Except the fall never came; I was locked in the moment of slippage. It was a less-intense sensation than pain, but equally destabilizing. I had a general sense of slippage as a finance-world term, something about expected outcomes not being met, but I didn't know anything about finance bros; I knew about earthquakes. A fault—the crack in the surface of the world where two tectonic plates rubbed against each other, sometimes causing earthquakes—

could move in varied ways, which in earth science were referred to as slips. There were dip-slip (vertical) faults and strike-slip (horizontal) faults, and varied angles and permutations of each. The Loma Prieta earthquake was an oblique slip, involving both horizontal and vertical types of movement. In all directions these two chunks of the lithosphere had pressed and pressed, and when the pressure became too much, they slipped, and the world broke. The pressures of life on the surface of that earth might be extraordinary or mundane. Something bad happens and a person slips and loses a job, or a car; the house goes next, someone gets sick, money gets tight, relationships crack. One slip led inevitably to another until it was impossible to stop the ineluctable slide into the indignities of misfortune; you were just like everyone else now, a member of the regular horde, a person whose circumstances are beyond their control. Collateral damage. Max saw it all the time at work. People got so used to being in crisis that eventually their baseline shifted. Like with a fascist president. Like weather forecasters on the news talking about record-breaking heat waves as though they were beach days. Like pain.

On the balcony my dad kept writing, true to his creative instinct even amid disruption, and I felt a strong desire to protect him from the slips and cracks of the coming era. To stand my ground. I knew it was unlikely, but I still wanted us to be excepted from the chaos coming our way. We were all tough cookies, my family. The fires wouldn't come for us, our jobs would be okay, we could pull it off if anyone could. We'd hang on to our loose interconnectedness, our doomsday humor, and our blue-sky beaches for the rest of our days.

· 3 ·

Glacier

Virginia Lakes
08.30.20

dear max,

my love. by now you have found and made camp. it's only 6pm but the sun is behind mount Olsen, elevation 11-thousand-something feet, and again behind Black Mountain, elevation 12-something, and again behind a cloud and i hope you are not too cold, my love, up and out there. it's funny that i count the nights until i see you again because i see you on so many of my nights but, well, i do.

this place is odd. quiet mostly but for trucks that pass. we are so close to Nevada and the air so dry. last night was a rough one, awoke first in a sweat down the back of my neck (a dream? fire fear or pain?) then next awoke to the sound of mice, i think, let's say it was mice, eating all my paper towels and leaving their shit in my bags as they passed through. i had the foresight to lock the food in the empty cooler but they still rummaged, so loud, seemingly everywhere once you know they're here. at 3:30am it was my own body that woke me, still not at rest. the day got better, i wrote and wrote and did not putter but did did did. the cabin reminds me of our log cabin, our first year in the redwoods, straight from a kit, the same kit no doubt, except there are three fisherman cots in the living room instead of our soft orange couch. it feels like one of those places we might visit off season in another country, plenty of guests but still a sensation of being abandoned or left behind. the lodge store is one of those shops that's 99 percent fishing stuff and the rest is candy with a visible coating of dust. the creek is audible at night; the chipmunks climb the screen door and beg like dogs. it's perfect, except you're not here.

i don't know what is happening out there in the world but i feel lucky to be heading towards it with you. when we kissed goodbye at the trailhead and i held your body, really felt your mass against my mass my hand at your back i wanted to tell you that i want all of it, feel all of it, the burning world, your pulse beneath skin, your you-ness of you. i'll hand you this piece of paper four days from now. sleep tight.

-m.

It had been eleven days since we evacuated. In the cabin I finished my un-sendable letter and took it out of the typewriter's scroll. I looked at my smartphone, expecting information; without internet or phone connectivity it was just a clock now, which made me hate it and love it at the same time. I passed time by making my dirtbag dinner—boxed mac 'n' cheese, with more plasticky cheddar grated on top, and a side of potato chips—and drinking my daily ration of beer. After the sun went down it became very dark. The light by the reading chair was harsh and reflected the sliding deck door in a way that made stark the difference between the expansive landscapes outside the door and the clumsy, if analog, technologies inside.

I wasn't getting much sleep. The mice were less cute than I described; they were ravenous, high-risk for hantavirus, and seemingly everywhere at once. Here too the smell of smoke was subtly present, but often in the daytime the sky was clear; the heavier smoke masses were locked in a holding pattern inside the Central Valley, west over the Sierra crest from here and at lower elevations. But every night in the small hours I woke coughing as some sort of shift in the air currents occurred. Wildfire smoke seeped through the dry pines outside, and the smell triggered new waves of fear. If a new fire came to Yosemite, how would Max, hiking alone in the backcountry, know about it? Would the rangers go and get him out? And if fire came to me here in the dry eastern highlands, how would I communicate to Max that I'd had to leave? It had been the right choice to come, I knew—I couldn't be at home alone, the anxiety would destroy me—and under less-precarious circumstances I loved a good, unplugged vacation, but no phone or internet also meant no fire news. Without information I worried about everything and everyone I loved.

I was writing letters to Max because I couldn't manage to write anything else. I happened to be working on the first draft of a novel that took place in the aftermath of a natural disaster, and I had come to the mountains with the intent of channeling some of what I was experiencing into pages. I had brought my old manual typewriter because there was no internet, and because the singular utility and percussive rhythm of that machine sent me back to a creative time in my life when I made culture as an act of pure expression, with no thought of it being a product. I had brought a week's worth of food, mostly beer and mac 'n' cheese, and the go bag, which I didn't want to leave at the house in case our fire, named Walbridge, which was still burning but still hadn't crossed the river, made any sudden turns. I had brought along a few books by authors whose language offered me joy during times when I couldn't focus on plot. Among them were works by Virginia Woolf, James Joyce, and Mary Ruefle. I recognized there was a good deal of hubris in the assumption that I'd be able to get writing done during a time of crisis, but I also knew that the crises seemed never to stop cascading; at some point, if I wanted to do something, I would have to do it under these circumstances.

In his polemic *The Great Derangement,* author Amitav Ghosh described the contemporary era as one in which the biggest, most obvious thing ever to happen on this planet—climate change—was, bizarrely, not the primary topic of most literature or art. Instead, culture offered what he called "modes of concealment" with which to further humanity's collective denial about the climate. Ghosh's argument, made in 2016, was that without cultural production that addressed the crisis—not science, not activism, which were also necessary, but *culture*—humans would keep failing to rise to the occasion and slow the irreversible effects of our collective addiction to burning fossil fuels. For authors of literary fiction, Ghosh said, to write about such a monolithic disaster felt like an impossible assignment. As I tried to make progress on my manuscript, plotlines seemed in danger of appearing ripped from bad blockbuster films. It's possible we creative artists just aren't up to the task, I thought. I put away the writing stuff and picked up a book.

The failure of literature to address the climate crisis, Ghosh believed, was at heart a failure of imagination. The scale of experience was too epic to envision; the scenarios too extreme to be rendered in story. Except it was entirely imaginable, I thought, because it was happening. The collapse of the climate was an unstoppable denouement of the Anthropocene, the era in which humans, through colony and capital, were the driving force behind changes in the geology of the planet. Its progress must have left traces in the literature earlier on. Maybe we just weren't reading closely enough. As I read the old faves I'd selected, I realized these books might be interpreted as disaster novels in their own ways. *Mrs. Dalloway* was a novel about class in which the characters are reeling from the aftereffects of unimaginable war, pandemic, and empire. In *Ulysses,* beneath all that language was a novel of grief set against the backdrop of a colonized country: there was Leopold Bloom just trying to do his day job and maintain his relationship; meanwhile, Stephen Dedalus, the Telemachus to Bloom's Odysseus, was being split apart by the twin deaths of his mother and his faith. For carrying along on hikes, I had brought Mary Ruefle's slim book of prose poetry, *My Private Property,* which described a more interior apocalypse: that of aging inside a body and the double-edged invisibility conferred upon women by infertility. Anything that ever happened to anyone had been unimaginable at one time, until it happened.

On my second full day at the fishing cabin I still couldn't write, and I found it difficult to concentrate on reading. After the adrenaline roller coaster of the prior week, my mind needed a break. I made tea and sat on the cabin's little deck and looked across the creek to a silvery granite peak. I needed to get closer, I thought, to touch rocks and shale slides with my feet, to see the dregs of the molten core of the planet once sheathed in ice. Like so many other tourists before me, I felt called to the mountain. I decided that for the next few days instead of writing

I would hike. I could write once my body inevitably felled me; today I would walk around in the most beautiful place on Earth.

31 aug 2020
lucky cabin #13, Virginia Lakes

dear Max,

i hiked today. up and around and over. lake, rocky pass, lake, lake. there was an old miner's cabin from the 1800s still standing, rusted tin cans displayed along with a century of lovers' initials deep in the wood. i almost carved our "M plus M" but decided it wouldn't be cool. i imagined the miner and his donkey, trudging up that rocky path, on a trail first made by Paiute peoples. one step at a time just like me. how he must have loved this place as he exploited it.

i thought about all the things in the mountains that might kill you in a cabin like that. i came across a young deer on the trail and we locked eyes, she wasn't scared though. at Cooney Lake (elevation 10,246) i ate nuts on a rock in the wind and read poetry and let myself feel the sun, so close. i would like to hike some distances with you, i would like to be strong and do that.

in the cabin next to me three men are staying, they look like they have money, they are motorcycle bros out on tour together. like me they do not fish, although that's what everyone else is here to do. they sit on the back deck overlooking the creek and don't really talk. i wonder about their home lives, do they also not talk to their wives, or do they just not know how to be alone with other men. but they are here. they are looking at the creek. they think it's beautiful too.

i looked on the map tonight to see where you are sleeping, Merced Lake. i hope it is not too windy and i hope you can see that same fat moon i can see tonight, only seventeen miles east of you, on this creek by this lake between these mountains at the other edge of California. i love you. Sleep tight.

-m.

p.s. i named the mouse Lightning.

The Sierra, the Sierra! Home to infinite granite-walled valleys, lush beflowered meadows, fabled giant Sequoias (cousin to the coast redwood), and proud peaks that topped out at fourteen thousand feet with Mount Whitney, the highest mountain in the lower forty-eight. One of the books I'd brought with me to the cabin and failed to open was John McPhee's *Assembling California,* part of his multivolume exploration of the geologic foundations of the United States. The book began with a humorous bit in which the author traveled west across the Nevada–California state line with a geologist in tow. The scientist had just given the author a tour of the geology of the Great Basin, and he tried to point out geologic features along the road. But as they crossed into California and entered the High Sierra, the rocks along the roadside started to get bigger and weirder, displaying complex layers of tumult in their striations. The Nevada geologist told McPhee he couldn't go any farther. This was not his wheelhouse, although it was just minutes away from his home turf. He was done. McPhee was going to have to get a California guy for this.

For California guys, the crown jewel of the Sierra Nevada was Yosemite, where Max was currently backpacking. Since Yosemite was declared a national park in 1890, its granite monuments and breathtaking waterfalls had become a worldwide symbol of California's natural beauty. Yosemite was the most famous part of the Sierra Nevada, but the range was about four hundred miles long, from the Tehachapi Pass in the south to the North Fork Feather River, north of Lake Tahoe.

The Sierra Nevada contained dozens of parks, national forests, and wilderness areas, including Sequoia National Park. On the western side the mountain range was green and lush; at the base of its piney foothills lay the Central Valley, which had formerly been floodplains filled with riparian forests and grasslands, and was now occupied by irrigated farms that grew a quarter of the nation's food. The Eastern Sierra Nevada, where I was staying, was less like a postcard and more like another planet, but it was no less beautiful. Here, in the arid shelter of a rain shadow that extended eastward through the Great Basin, the peaks grew sharper, the wildflowers smaller, and the rocks less forgiving. Just over the next ridge from me was the salty expanse of Mono Lake, and beyond stretched the desert regions that author Mary Austin had nicknamed "the land of little rain."

Fire and water had shaped this place. As McPhee's next geologist, a Californian, showed him, the Sierra Nevada's superlative granite monuments, vertiginous valleys, and craggy mountaintops were the result of eons of geologic upheaval. Way back in deep time, over the course of hundreds of millions of years and many climatic shifts, these landscape features were created by the movement of magma, tectonic plates, water, and ice. Beginning in the late Jurassic and Cretaceous eras, earth and fire had initiated the process. At the time this was the edge of the continent, one of many cracks in the surface of the world where two tectonic plates met. As the eastward-moving Farallon Plate subducted beneath the westward-moving North American Plate, molten rock intruded into the upper levels of the earth's crust. The magma cooled slowly over time, resulting in the distinctive granitic rock of the Sierra Nevada batholith. Meanwhile, aboveground, volcanoes were fed by more magma that erupted as lava, cooled quicker in the open air, and layered additional rock and mineral deposits above the granites. Towering monoliths such as El Capitan and Half Dome were at one point underground.

Water soon joined fire and earth to play its role in shaping the land: depending on which geologist you ask, either three million years ago or forty million years ago, give or take, the Sierra Nevada range began to uplift. This major tectonic event also created massive rivers that carved the landscape into steep canyons. In the ensuing epochs the mountains were covered by many series of glaciers, which in turn sculpted the land as they moved, eroding layers of rock and other deposits and exposing the ancient, underlying granite. It was ice that carved the distinctive U-shape of the Yosemite Valley, while freeze-thaw cycles of precipitation shaped the signature jagged peaks around it by causing cracks in the rocks to expand and burst. Eventually, all that water had to go somewhere: glacial melt and megarivers became gargantuan torrents that flowed west to the Pacific Ocean, ultimately forming the valleys and watersheds and bays of the land now referred to as California.

Yosemite was rarely spoken of without reference to conservationist John Muir, but when he arrived there the Sierra range was already occupied by Indigenous nations. The Yosemite Valley has been the home of the Ahwahnechee people since before history was written. Gold miners were the first non-Native people to forcibly settle the area; Muir arrived in 1863 to moonlight as a shepherd and study glaciers. He ended up founding the Sierra Club and advocating for the preservation of Yosemite as a park. In the mountains Muir encountered a splendorous landscape unlike any he'd ever seen; he famously called it "Nature's design." But the people who already lived in nature knew differently. The pines, manzanita, chaparral, and grasslands Muir waxed rhapsodic about in his journals didn't just magically grow into arrangements that inflamed visitors' aesthetic passion. As ecologist M. Kat Anderson wrote in *Tending the Wild*, a history of Indigenous land management in California, what looked like wilderness to colonizers was in fact "a carefully tended garden that was the result of thousands of years of selective harvesting, tilling, burning, pruning, sowing, weeding, and transplanting." The Indigenous people managing this garden were the same people Muir was describing in his journals as "dirty" and "belonging to another species."

Contrary to what I had been taught in elementary school, Indigenous people in pre-colonization California didn't merely wander around passively receiving the bounty of the land. Over millennia Indigenous societies developed and passed down methods of cultivating the ecosystems in which they lived that allowed the land to function as a resource while also sustaining the intricate network of systems that made it useful in the first place. Gardening the landscape was how communities got what they needed to survive and—crucially—how they ensured continued survival for all of an ecosystem's residents, human and more-than-human. The riches of ecosystems were cared for and coaxed the same way a kitchen gardener might care for her vegetable patch, or the way I tended my apple trees.

In many parts of California, including where I lived, a gardened landscape might look like a stand of oak trees: acorns were an essential food and cultural resource for many people and animals; the introduction of regular, moderate fire regimes encouraged oaks to produce more and healthier crops more frequently. A garden could also be a shrub: the intentional burning of hazelnut bushes forced the growth of young, strong, straight shoots, which were ideal for making baskets or other vessels. Muir himself observed many instances of Indigenous people using fire to create feeding grounds for the deer they hunted. Fire had been—and still was—a primary tool of Indigenous land management, but the tool kit was diverse. Indigenous stewardship included long-term crop planning; Anderson cited a Chukchansi elder named Clara Jones Sargosa describing how harvest laid the groundwork for future seasons: "In digging wild potatoes, we never take the mother plant. We just select the babies that have no flowers. . . . We are thinning the area out so that more will grow there next year." Such interventions benefited larger food chains and systems in addition to human-involved ones. In the present day some Indigenous people, mostly outside of urban areas, still relied on traditional crops as food and cultural resources, and still more were organizing and agitating for their rights to do so.

As these uses demonstrated, Native land management had always included specific practices, which—historically and currently—varied between different nations and ecosystems. In the 2018 anthology *Traditional Ecological Knowledge: Learning from Indigenous Practices for Environmental Sustainability*, editors Melissa Nelson and Dan Shilling, both Indigenous themselves, argued that a common thread of Indigenous land management is the incorporation of an ethic, not merely a list of tasks. For example, with their hunting skills, in precolonization days "American Indians probably *could* have wiped out the bison, but they didn't, an act of ecological restraint and spiritual reverence," wrote Shilling.

Inversely, since the 1850s colonialist settlers had mostly revered the veins of gold and silver that ran through the hills and foothills of the Sierra Nevada. Mining was the primary extractive industry in the Sierra and an industry around which others, such as logging, were centered. Early conservationists like Muir campaigned to prevent the decimation of the land by mining companies. They largely succeeded; since the early twentieth century, tourism had replaced mining as the main extractive economy in the Sierra Nevada.

The history of the world was so often the history of men not knowing how to respond to beauty. When settlers first encountered the magnificent landscapes of the High Sierra they knew instantly that they were in the presence of a superlative natural power. Some, like the miners, reacted by wanting to tap it, exploiting the land's natural resources for profit. Others seized on aesthetic power; Muir's journals are packed with ardent descriptions of the landscape that at times read more like professions of romantic love than like field notes. Conservationists and extractive capitalists may have appeared to hold opposing views about nature, but both ultimately saw landscapes as ripe for possession. Early conservationists like Muir had fallen in love with the Sierra, and that romance had indirectly led to my own love of the place, but it was a passion that disturbed me when I read their writings. It was less like love and more like an obsessive crush: overly romanticized, objectifying, and potentially dangerous. Whether for pleasure or profit, such men wanted to reserve for themselves the places they designated as wilderness, to make of these mountains man's playground. Ideas of purity and exclusion were at the heart of both approaches. Not coincidentally, both groups viewed nature as inherently feminine. And both groups' visions of the natural world included the subjugation and exclusion of people of color, Indigenous people, poor people, and people with disabilities—all of whom were deemed undesirable or unproductive when it came to the needs of capitalism or recreation.

In its heyday the conservation movement had close, and well-documented, ties to the eugenics movement, which advocated for the control of human reproduction in order to create a white master race. While conservationists invoked a message of romantic innocence to achieve their goals, eugenics cloaked itself in a language of science and medicine; its theories were later taken up by Nazi scientists. Earlier in summer 2020, as protests across the United States had inspired the latest public reckoning with systemic racism, the Sierra Club released a remarkable statement highlighting and denouncing its exclusionary legacy. One of many examples of the links between early conservation and racist ideologies was the work of David Starr Jordan, a member of the Sierra Club board during Muir's presidency (and also the first president of Stanford University). Jordan founded multiple explicitly racist organizations and was one of the most prominent eugenicists in the nation. He advocated for, among other horrific policies, the forced sterilization of women deemed undesirable or unproductive in eugenics' imagined "pure" society.

These two driving environmental philosophies of settler colonialism—ripe for profit or pure for plundering—occupied what Kat Anderson called "flip sides of the coin of alienation." What they had in common was a sense of exclusion that rendered nature "an abstraction," Anderson wrote. Whether a person wanted to deflower nature or put it on a pedestal, such desires were supported by the same premise: that humans were not members of the natural world, and nature was something we should—and could—control. Naturally, controlling the land included the exclusion of anything those in power perceived as ugly or unprofitable—and fire was the ugliest, most destructive monster that a man possessed of a landscape could conjure.

The Sierra Nevada had long been a fire place as well as a place of ice. But when the Indigenous populations' use of fire to manage the land became verboten, the Sierra forests began to be overcrowded. In his 1950s research, published as *Forgotten Fires: Native Americans and the Transient Wilderness,* anthropologist Omer C. Stewart compared a picture of the Yosemite Valley from 1866 to the same view in 1961; the forest in the 1800s was remarkably sparse. There were fewer trees overall, more space between trees, and higher diversity of species. The present-day forest, despite its infrastructure and traffic, was the true wilderness. It was wild because it was neglected: a garden left to ruin.

Earlier that week, when Max and I arrived in Yosemite, we had stopped at the park entrance to register Max's permit, and at the ranger station a friendly sign let us know the locations of all the fires currently burning within the park. I began to drive the sixty-five-mile route across Tioga Pass to drop off Max at his trailhead. Soon we saw live fire just down the hill, off the road. A helicopter flew by, close to where we were at nine thousand feet elevation, and that familiar fear rappelled down my spine again. But we soon passed another park sign that politely asked us not to call 911: it was a lightning-sparked fire that was being monitored by the park but allowed to burn out. In the 1970s, Yosemite began again to experiment with more aggressive test programs of prescribed burning and a policy of allowing wildfires more leeway when they did start. The contrast in my experiences of fire in Yosemite versus at home had been striking: large areas of the state were in a flurry of evacuation, with tanker jets and armadas of personnel trying to quell unstoppable flames. But here, the park management seemed to be doing what they now knew everyone should have been doing all this time: tending the land by letting fire do its job.

One of the paradoxes of the wildfire crisis was that *good fire*—intentional, managed, low-intensity burns set by humans—could help prevent harmful wildfires. As we in California were experiencing, landscapes deprived of good fire were more likely to experience severe damage in the event of a wildfire. While good fire wasn't a panacea, it was a powerful tool. As wildfires had become hotter, bigger, and more frequent over the past decade or so, attitudes had begun to shift away from the Smokey Bear paradigm, but there were 150 years of neglect to make up for in a garden the size of the American West. In addition to the forests I so loved, which made up a third of California's land, grasslands, prairies, chaparral, and many other ecosystems were evolved to be fire-reliant. In 2023, the U.S. Forest Service estimated that to restore balance in these fire-suppressed landscapes, they would need to preventatively treat twenty million acres of California's federally owned forest- and grassland in the next decade. "Treatment" included beneficial fire but also the clearing of dead and dry fuels, infrastructure work to restore healthy water flow, and the thinning of young trees. (Commercial logging, although technically an act of tree thinning, increased wildfire risk because in the timber business, bigger was better: logging harvested the oldest, strongest trees and left gaggles of young saplings to serve as kindling for wildfires.)

When it came to prescribed fire specifically, the fire industry newsletter *Wildfire Today* estimated that between 2017 and 2020, about 120,000 acres in California had been treated with beneficial fire; a 2021 report from the state's Wildfire and Forest Resilience Task Force set an "aspirational" goal of burning 400,000 acres by the year 2025. By contrast, it had been estimated that before colonization, 4 to 6 million acres of land burned annually in the state, a mix of naturally occurring and anthropogenic fire. Paradoxically, the fire season I was currently experiencing would ultimately match that number, burning over 4 million acres. For land, people, and climate, the differences were intensity and harm. Fire scientists and archaeologists were working hard to learn how much historical burning was intentional, but it was widely acknowledged that the portion was significant. There had always been massive wildfires in California, but regular human use of low-intensity fire with specific goals had helped contain them. Along with Indigenous fire management, the annual frequency of low- to mid-intensity wildfires was itself a historical factor in preventing potentially catastrophic events. The scale of need was staggering by any standard.

Restoring good fire wasn't simply a matter of acres burned. There was also a need for a fundamental shift in the dominant culture's attitude toward fire and nature. When I discussed the "20 million acres" goal with Andrea Bustos, a prescribed fire specialist living and working in Northern California, she offered wider perspective. Bustos is from Ecuador, where she created groundbreaking integrated fire management programs with rural and Indigenous communities. It was she who first introduced me to the term *neopyrocolonialism,* which she loosely defined as fire management practices and institutions that replicate the behaviors and harms of colonialism. The Escuelas de Campo and integrated fire management trainings she developed in Ecuador were examples of countering that mindset. The model was simple: ask local communities what they need to survive, and give it to them. After all, they were the experts in protecting their own environments. In Andean worldviews, Bustos told me, fire is an integrated part of everything, an element of life. So when it comes to prescribed fire, she said, It's not like we burn X acres and then the bad fires stop and we're all good. Good fire is forever. Good fire requires changing who you think you are and what you believe.

As Bustos indicated, in addition to huge shifts in resources, labor, and lifestyle, for institutions to implement cultural and prescribed fire at the scale needed, a more reciprocal relationship with the natural world would be necessary. To shift back toward balanced land management practices would require a system-wide, culture-wide, and deeply person-specific adjustment in people's relationship to the land. To burn intentionally at such scale would require conceding that the Muirian ideal of wilderness was an invention. Idylls of escape and leisure would shatter, the power of fire ever present. For many it would also mean the loss of a fantasy: an acknowledgment that although humans could have a large-scale effect on nature, we couldn't actually control it. If people not native to this land allowed fire to return, we would have to allow for the fact that the land was not ours.

A future memory. It's a couple of years after the lightning fires and I am in Lake County, which borders northern Sonoma County to the east. If Sonoma County is considered high-risk for wildfire, then Lake County is a sure bet. I'm standing with about two dozen people on the shore of Clear Lake, the largest freshwater lake in California and the oldest lake in North America. Around us, mountains are visible on all sides. Nearby, a row of empty-looking vacation houses looks out at the water. At the edge of the shore, reeds grow. It's springtime and it's gorgeous here, but beyond my sight line, on the shore opposite from where we stand, there is a 160-acre Superfund site, the former open-pit Sulphur Bank Mercury Mine. The company began mining mercury and sulfur here in 1865 and operated until 1957, polluting the food chain with mercury and arsenic the whole time. This gorgeous place has been poisoned. Seven Indigenous nations are based around this lake.

I am here to observe an exchange of ecological knowledge between a group of Indigenous Americans—Pomo citizens—and a group of Indigenous Mexican people, all immigrants to Sonoma County who worked the land in their home countries and now do farmwork here. The locals are part of Tribal EcoRestoration Alliance (TERA), a cross-cultural alliance that trains and employs people in land management and fire resilience work, with the goal of uplifting Indigenous perspectives and leadership. The visitors are farmworkers organizing for better conditions as they labor on the front lines of climate change.

(I'm able to observe this exchange because the labor nonprofit where Max now works is supporting the farmworker campaign, and Max wrangled me an invite.)

After an official welcome speech from a Pomo elder, the TERA crew leader, Stoney Timmons—ponytail, beard, trucker hat—explains that the reeds we are standing in used to grow around the entire lake. Tule, pronounced *two-lee* and sometimes called bullrush, is a tall, strong reed that grows along waterways. Tule is a staple of survival in our Pomo culture, Stoney says. Its fiber can be used to insulate housing structures, build boats, or weave baskets. It is also food, he adds: get the roots fresh, roast them, and add a little salt. It's tasty.

Stoney shows us three different sections of tule on the shore. In an area that's been left unmanaged, the vegetation is crowded and dry. It's a fire risk. Dead stalks shade out new, green shoots. Birds are easily camouflaged in this scrim, Stoney tells us, which causes them to overhunt hitch, the small minnows that have always been a staple food for Indigenous people in the area. Next we look at an adjacent area of tule that burned in the 2020 wildfires. Stoney picks up a chunk of burned dirt and turns it over. The soil is about six inches deep; its underside is a tangle of charred organic matter. Tule is a rhizomal spreader—the plants have a reserve of corms, or tubers, from which the root system grows, allowing tule to reproduce and rebound after fire. The extreme wildfire killed these tubers, ending the life cycle prematurely. Stoney then directs us to a smaller patch of land where he and his colleagues did a prescribed burn earlier this year. After good fire, the tule reeds are growing well. They are thick and strong, and because the shoreline isn't overgrown with hiding spots for birds, animal populations are more balanced. (You still can't eat the fish, Stoney explains, because of the mercury.) This area is evidence of the effectiveness of intentional burns: it is far less likely to re-burn than the overgrown, dried-out tule we first observed, and if it does burn, it will do so at a lower intensity than the wildfire-killed tract.

When Indigenous people use fire it is often described as cultural burning or cultural fire. The phrasing reflects the integral nature of fire to life. In diverse Indigenous worldviews, fire's significance as tool, medicine, and spirit is impossible to separate from fire's ecological importance. Culture is land, land is spirituality; spirituality that values a frictionless and reciprocal existence with the more-than-human world is also survival. Cultural fire could describe fire used for heat, cooking, ceremony, land management, hunting, or warfare. Cultural practice is specific to each person, family, and community. Depending on the practitioners, situation, and scale, cultural fire might be done using traditional methods or modern technologies. When I ask Stoney if cultural burns include a prayer or traditional offering, he says, Yeah, but people often get hung up on that aspect—and by "people" I take him to mean journalists and white folks, most of whom, like me, grew up ingesting cultural stereotypes of *magical Indians.*

Some people think you have to go out and say all the prayers and stuff, Stoney tells me, but when we're working, our sweat is also an offering. He continues: Just to be out there and feel the land is the biggest offering I think I could give.

Stoney knew ceremonial fire growing up, but he first learned to put fire on the land when he joined TERA in 2020 and attended a TREX—Prescribed Fire Training Exchange—an annual series of trainings and burns hosted by different fire organizations. The TREX was hosted by the Yurok nation, up near the Klamath and Trinity Rivers in Northern California. The Yurok, along with their neighbors the Karuk and Hupa, are widely considered leaders in the practice of fire. After his first burn, Stoney was hooked. Fire, he says, is a healer.

The training that Stoney attended was led by Margo Robbins, an enrolled member of the Yurok Tribe who is a fire expert and educator. Robbins is cofounder and executive director of the Cultural Fire Management Council, an organization working to bring fire back to Yurok ancestral lands. The good-fire world is small: after my visit to Clear Lake, I end up interviewing Robbins about the difference between cultural and prescribed fire. The important thing to understand, Robbins says, is that when we're burning for cultural purposes, a byproduct of that is wildfire prevention. When agencies are doing prescribed burns, they are burning for fuel reduction, and cultural resources might be a byproduct of that.

The subtle difference in intention that Robbins describes reminds me of what Andrea Bustos said about the worldview that good fire requires. The people at TERA appear to be trying to bridge the gap between cultural and prescribed burning; when they do prescribed burns, one TERA staffer told me, they require inclusion of cultural goals in the official burn plan. For example, a fire or forestry agency's prescribed burn plan for an oak woodland might include a goal of reducing a certain percentage of Douglas fir saplings, as firs are fire-vulnerable and crowd out fire-adapted oaks. A cultural burn goal for the same area might be to burn for the consumption of leaf litter and duff on the forest floor, in order to reduce habitat for acorn weevils. Reducing the pest habitat improves the quality and quantity of the acorns, which means there will be more acorns for people to gather and eat, more seeds for animals to store and bury in the ground, and more future oak trees for the land.

At Clear Lake, Martin Duncan, TERA's crew safety coordinator, walks a small group of us through the tule. Marty is about my age; he wears a hoodie and I can see forearm tattoos peeking out where he has one sleeve pushed up. Tule is almost like water, he says, it's life. He then tells us a story about Lucy Moore, his great-great-grandmother. In 1850, when Lucy was a small child, she survived a massacre. Two white settlers—Charles Stone and Andrew Kelsey, after whom the nearby town of Kelseyville was named—had been enslaving Indigenous workers and violently abusing them. The workers rebelled, killing the two men. Kelsey's brother summoned the cavalry, then made up of both soldiers and volunteers, who rode to Clear Lake looking for the workers. Instead, they found Pomo citizens, including children, gathering food and plant medicine on Bo-No-Po-Ti, a small island in the lake. The cavalry had bayonets. They didn't stop using them until everyone was dead. It is thought that between fifty and two hundred people were murdered that day on the island that became known as Bloody Island. Lucy was there. When the massacre began, she jumped in the water to hide. While submerged, she grabbed a reed of tule. Martin gestures to the plants around us. He says, This tule is literally the reason I'm alive today. Tule is tubular and hollow; Lucy, as a child in a community that valued ecological knowledge, knew this. She bit into the reed to break it open, placed one end in her mouth, and used it as a snorkel. In this way she breathed, hiding underwater, until the water around her was red and the soldiers had gone. (The mob headed west, looking for the men who'd killed Kelsey, and ended up murdering another seventy-five Pomo people in the Russian River area, where I now live.)

After Martin relates this history, a woman named Ana, who is with the group of immigrant workers, pulls him aside. Ana is short and wears thick black eyeliner; a community leader, she has a way of speaking that is low-key but commands attention. She says to Marty, I was moved by the story of your ancestor; I think we share many things. She tells him her own story: Ana crossed from Mexico into the United States thirty years earlier, when she was twenty-six years old. The coyote—a smuggler paid to escort people across the border—abandoned her group in the desert. They wandered for three or four days without food or water. Thirsty and overheated, Ana began hallucinating. Another woman in her party died of exposure in front of her. Ana was certain she would die, too, and she saw in her mind's eye a vision of her grandmother Maria, who had passed when Ana was a girl. Maria, an Indigenous Mexican woman who spoke Nahuatl, had been a healer and midwife. When Ana was a girl, Maria taught her that a certain type of cactus could store water in its roots in little sacs. (Later, when I ask her if I can write about her story, Ana recalls playing with the sacs as a child, throwing them like balloons, and Maria scolding her for disrespecting Mother Nature.) In the desert that day, delirious, Ana saw her grandmother, and her grandmother said: The cactus. Ana and her companions soon located one. They dug deep underneath its roots, found the small sacs of water held there, and drank them. Around the cactus they also discovered tiny frogs—their skin was clear, Ana recalls—who were drawn to the same water; the travelers ate those, too. When she got out of the desert, Ana wondered whether the experience had been a dream. Then she looked at her hands: her fingers were torn and bloody from digging. At this point in Ana and Martin's conversation, I can see that Ana is crying a bit and I back farther away, wanting to give them privacy. I am still within earshot when she says, The plant saved my life. Martin, quiet, nods.

When I had booked my dirtbag evacuation week in the Sierra, reservations had been difficult to find on short notice, and I had ended up with two stays at cabins in two different canyons. My last morning at the first cabin, I saw the mouse, or one of them. I had taken to carefully cleaning the surfaces of the cabin's kitchen every morning because of the infestation. After I packed up, I swept my sponge across the tinfoil-topped stove a final time, and a piece of last night's macaroni fell to the floor in the crack between the appliance and the wall. I looked down and there she was—little Lightning, one of what was no doubt a voluminous squadron of chaos-makers who had rendered my nights intolerable. The mouse was cute and crouched on her hind legs. She didn't run when she saw me. In the way that people have of imparting meaning onto encounters with other animals, I liked to imagine that we locked eyes, as I had with a deer near the miner's cabin the previous day. In that moment, perceiving this minuscule life-form, so flawlessly evolved to survive, my wakeful nights and cleaning sprees seemed ridiculous. What was I doing, hiding and locking away my food from this critter? I was a visitor in her house, just passing through. I had my own home, full of crumbs and luxuries, which last I'd heard still existed and hopefully would continue to do so. I batted several more macaroni noodles off the stovetop and let them fall to the floor, unswept. I watched as Lightning grabbed one in her little paws and retreated to the netherworlds of the cabin. Good luck, buddy, I said, and I loaded my typewriter into the car.

On foot from the eastern edge of Yosemite a hiker could walk over 11,000-foot-high Parker Pass, take a steep descent named Bloody Canyon (I didn't want to know why), exit the park bounds, and end up at the cratered shores of Mono Lake, elevation 6,378 feet. My first cabin had been near to this eastern border of the park, so I approximated the Bloody Canyon approach to the lake in my car. The basin of Mono Lake rose into view and its surface cast smokelight back at the sky. I saw a sign for a county park on the lakeshore, pulled over, and masked up. A nondescript marker near a low concrete shed read PROPERTY OF THE CITY OF LOS ANGELES. Past an empty parking lot and a low building containing public restrooms, a boardwalk led toward the water. It was Kutzadika'a land.

The Sierra Nevada mountains were made by fire and water, and they also made water: by the year 2020, the Sierra snowpack provided 30 percent of the drinking water in California. Hetch Hetchy reservoir, a dammed part of the Tuolumne River, which began in Yosemite, provided municipal water for the San Francisco Bay Area. Since the forties, Mono Lake's feeder streams had been diverted to hydrate the city of Los Angeles some 350 miles away. If water had shaped the first half-billion years of life in California, it was lack of water that would shape the coming era. The Sierra snowpack was not exempt from extreme weather. The dry seasons were getting longer, disrupting the snowpack's cycles of accumulation and melt. Most parts of the state had been in a drought since 2011, dipping briefly out from time to time when the rainy season brought short bouts of extreme precipitation and floods before again plunging back into water precarity. The Bay Area would soon careen back into drought level D4—*Exceptional,* the highest level of drought—before record rain whiplashed us back out and the pendulum swung again. The drought category D3, *Extreme,* was in part defined by its ability to extend the fire season into a year-round event.

The city of Los Angeles began diverting Mono Lake's feeder streams for its own use in 1941. The lake lost 68 percent of its volume in the next twenty years. Mono Lake was a drying-out lake, a rich ecosystem that was in the slow-motion act of being forced to adapt. Millions of birds migrated through each season, and more than three hundred species lived at the lake year-round. Endemic brine shrimp, algae, and alkali flies provided the basis of the ecosystem's food chain. When the water was taken, the chain was broken. Partially thanks to all those migrating birds, an organized group of environmentalists and residents called the Mono Lake Committee won a major court victory in the 1990s that limited the amount of water L.A. could take. Saving the lake had slowed the decline, but its health was still precarious.

I walked down the boardwalk path surrounded by tall reeds and the odd Seussian structures called tufa towers. As a terminal lake with no outlet, Mono had high levels of salinity. When high-calcium freshwater rose from springs beneath the lake bed, it came into contact with the salty carbonite water, and a chemical reaction occurred. This reaction, of the same type that formed limestone, created tufa, stacked formations of rocklike calcium carbonate deposits. The tufa at Mono were geological rarities, sometimes thousands of years old. As the draining of the lake had lowered the water levels, tufa had surfaced and become terrestrial features as well as submerged ones.

Every few yards I passed a marker displaying the lake's levels in 1944, 1967, 1989, a march of terrible diminishment that ceased at the current all-time low: the actual shore. There, there was water. It was blue. There were birds. Tufa like monoliths jutted up from the lake. At the shore, I learned from a sign that one of the islands in the lake was the nesting place of the same California Gulls I'd seen on the beach in Santa Cruz. The low water levels had opened their island to land predators, and the nesting grounds, one of the largest in the world, were in danger of being wiped out. A few gulls had broken the ancestral cycle of migration and begun to nest closer to home, on small islands in the San Francisco Bay. The gulls in Mono were still declining but the ones in the Bay were thriving. They were now driving out other threatened waterbirds. In the process of shifting their migration patterns in response to human destruction of their habitat, the gulls too had become an invasive species.

From the boardwalk, the lake shimmered with midday heat. Salty mud bubbled imperceptibly around its rim. The magnificent, troubled machinery of the planet ground on. Birds called. Water licked the shore and the inland tufa stood stranded, parched. I pointed the car back to the mountains. This lake didn't seem to me like a place one should stay very long, even when one appreciated its resilience. It seemed to me then like a place that was dying, and it might take me with it if I stuck around.

i think it's Tuesday?
definitely 2020
Rock Creek

Max,

I am too tired to write you a letter tonight. Sleep tight, somewhere near Vogelsang, my love.

My last hike in the garden of the Sierra was supposed to be short, although the trail continued for days and nights if I wanted it to. My second cabin stay was up a steep canyon draped in old burn scars, glittering aspen, and crackly pine trees. In the morning I drove a mile farther up the road until it dead-ended at a trailhead, and I entered the eastern side of the John Muir Wilderness. I'd do a short out-and-back hike, make a few lakes, and have lunch back at the cabin. Then maybe get some typing done.

The trail started slow: dirt and rocks, a creek-side ramble through a dry dip in the mountains. After a steeper-than-expected climb, I descended to a meadow: green grasses, trickling streams, and wildflowers backgrounded by epic rocky mountains. I forded a small stream and in the middle I paused on a rock, balancing with my hiking poles dug into the shallow water. I listened. Water. Life. I breathed. I was glad all those dead racists had preserved this place. It *was* sexy.

A truism of the mountains: the best lake is always the next one. From the pinnacle of each pass, the descent wasn't always visible, but

I was confident knowing what I'd encounter: Up and down, and here was yet another crystalline oasis; up and down again and a magnificent alpine vista. At the most turquoise lake of them all, I stopped to watch the wind make small ceaseless waves on the water. Their rhythm was so unlike that of the ocean, in which waves rose and fell in distinct sets as the tide changed. These inland currents were made of wind and ice, and their pattern never wavered; a lake's ripple was perpetual.

A MEMORY. I am nine years old and taking my first multiday backpacking trip in these same mountains. It is the mid-1980s; backpacks have external metal frames. The party is made up of myself, my dad, and my older brother, Niranjan, with my new stepmother, Stephanie, acting as guide. We begin the trip at the home of a woman named Jan. Jan is an older friend of Stephanie's; they met through volunteer work with the Mono Lake Committee. She's only in her mid-fifties, but to me she appears both ancient and timeless. I'm never clear on the details, but somehow Jan owns land along a small creek that flows into Mono Lake. In memory, everything is a bit blurry, my perspective tinged by childhood's soft focus. It is spring, which in the mountains probably means it is June. Tioga Pass is freshly plowed and open. On Jan's land the grass is new green. There is a cabin, a woodstove, a black cat. Jan has silver hair and wears wool sweaters like a proper mountain woman. Each winter, when the roads are snowed in, she has to ski out for supplies. My family and I leave our tents in our backpacks and sleep out under the stars, and dew seeps into my mummy-style sleeping bag and binds me to the ground. I wiggle so much in my sleep that when I wake to the rising sun across the lake, I've nearly inchwormed myself off an embankment and into the creek.

It is on this trip that I first jump into a glacial lake. At each body of water we cross, my dad offers a dollar to whomever can stay in longest, up to a minute (beyond that, he tells us, hypothermia will set in, although in truth the water would probably take at least an hour to kill me). I win the dollar every time. Not coincidentally, it is also this trip where I first learn to move through physical discomfort, to be a toughie, even when it hurts. There is a story Stephanie likes to tell to her third-grade students about this trip, about events I barely remember. The way Steph tells it, the days of hiking are long and feature the usual discomforts of the outdoors. My brother whines and complains the whole way, but I, a trouper, never say a word. I keep myself distracted by talking to an imaginary pet horse, and I put my head down and let the horse lead the way. In subsequent decades Stephanie often pulls this story out when she wants to compliment me, my quiet endurance, my imaginative pluck. In my memory, every step of the hike hurt. However, what I carry with me from that trip isn't an experience of silent suffering. It's the way glacier water punches my skin when I jump into it, naked; how the body adjusts at first slowly, then very quickly, to icy pain.

I FOLLOWED MUIR'S TRAIL until I reached Long Lake—or was it Chickenfoot Lake, the trailhead map that I'd taken a bad picture of with my phone was confusing. At the lake, I selected a spot on an incongruous patch of clear sand near a dip in the shoreline. Sitting low on a half-submerged rock, I unwrapped my cramped feet from their hiking shoes, unpeeled my sweaty socks, and plunged my feet in the water. At an elevation of more than ten thousand feet, the water was sharp. That shock, the ease into it, the deep pleasure of hydration, permeated my lowland body. This is where water comes from, I thought. I looked across the lake to where the horizon dropped into yet another canyon, and I felt not merely refreshed, I felt powerful. Despite the fact there were other hikers visible in my sight line at that moment, I felt as though I was an explorer in uncharted territory. I couldn't deny that when I stood and surveyed the terrain before me, I fancied myself the discoverer of it. I was alarmed to realize this feeling was not unlike the one I imagined that long-ago miner must have felt when he found his mother lode. The colonizer inside me wanted dibs on the lake.

The sun was already low behind canyon walls. The smell of evening smoke arrived on schedule. I'd hiked farther than planned, and I only had snacks and no lunch, but I was well stocked for water and I felt safe; it was high season, the trail was downright crowded, the faces I met on it appeared unthreatened by and uninterested in me. It was still a time in the Covid pandemic when going outside felt dangerous. I wore a bandana around my neck and pulled it over my mouth and nose when I passed another hiker, and I silently feared their respirations when they didn't do the same. I wanted to keep going on the trail, to discover the world for the first time over and over again, but I knew it was time to turn around. I was an amateur hiker but I understood the most important rule of the mountains, which was *don't push your luck*. The mountains can kill you without even noticing, or they can liberate your soul. Sometimes both.

On my return leg, I passed a father-son hiking duo for probably the fourth time; we'd been pacing each other all day. The father looked like he was having a hard time with his body; the son looked miserable, like he'd rather be anywhere else. They were both white, both ruddy. They wore similar T-shirts, not matching exactly, but the shirts looked like they'd been ordered from the same place: American flags with blue and black stripes inserted into the pattern, large eagles flashing their claws, and aggressive slogans with a strong right-wing nationalist bent. I wondered what this landscape meant to these men. Did they love its beauty as I did? Did they fear its power? Were they just doing something they thought dudes were supposed to do on vacation? In the subsequent century since Muir had hiked here, the outdoors industry had developed his verdant crush on the landscape into a consumer culture of what poet and nature writer Kathleen Jamie once described as the "lone, enraptured male." In this paradigm, a guy rambles around the mountains, finds awe and inspiration in their splendor and power, and thinks he's the first one. It must be exhausting. Although I supposed I wasn't that different from those men. I didn't know what to do with all this beauty, either, where to put it, how to process the conflicting experiences my body and heart had when I was outdoors: I knew that the concept of wilderness was a construct, and I felt how this wild place was sacred.

I stood aside and waited for the grumpy hiker family to pass. After they did, I found a good rock just off the trail and sat on it. My pelvis took to the granite with grace. During the hikes I'd been on that week, something unexpected had happened: in the garden of the mountains, my body felt good. I'd adjusted beautifully to the elevation. I felt energized, not exhausted, for the first time since the lightning. In my mind I traveled down into the nerve box of my pain to see what was going on in my body. I felt the glorious sensation of nothing. Beneath the pulse of my physical exertion I could sense the presence of an underlying waver, hints of the usual hurt, but today it was more an echo than a wail. There was less weight to my damage up here, in the sky.

Whenever I told people that my IUD attacked my body, there was a certain set of women—mostly Californians, mostly white women who came of reproductive age in the 1960s and whose tastes trended toward turquoise jewelry and Tibetan prayer flags—who responded by saying something like, Oh, yes, well, it's not surprising. IUDs, you know. Putting foreign objects inside the body never ends well. Their reaction was understandable; they remembered the Dalkon Shield, a popular IUD that was pulled from the market in 1974 because it caused sepsis, injury, miscarriage, infertility, and death. But I perceived in this maxim an underlying morality that made me bristle. I had never had surgery before the onset of my IUD-related calamities. I, like many holistic-leaning people, had an image of my body as a natural thing, a system that functioned best when left alone by the hands of the patriarchal medical system. But the problem with the idea of untouched nature is that nothing ever is. If bodily meddling was categorically wrong, where did that leave me when it came to using a foreign object to intervene on my body's natural tendency to get pregnant? Biological essentialism didn't work for women, and the idea of a pure, unmanipulated natural world hadn't worked for humans. Both relied on a fantasy, a story of innocence that had never been true. An intervention was not inherently good or bad; it was part of a dialogue. The tending of a natural body required constant attention, the giving and receiving of nurture and discipline. Extraction and tourism were types of attention. So was gardening. It mattered how the relationship was structured, not just that there was one.

I stood again and readied my hiking poles. Maybe, I thought, the wild wasn't something I needed to spend too much time fretting over. There were wildernesses I'd already visited inside myself, the twist of my spine when the pain lightninged through me. The deep, wet black that my pupils became when I inhabited my body as it was:

damaged, adapting. Maybe the true measure of wilderness was its inescapability. There was no wilderness like the wilderness of a body. No majestic landscape could mirror the nerve channels that were laced all through me. The horizons of exploration weren't necessarily *out there*, or even *down there* buried in rocks since deep time. What more adventure did I want than the one I was undertaking already, that of relearning the way my body now worked, what it would permit and what sensations thrilled it, for the first time, again.

A few nights after my hysterectomy, I had woken Max because I was howling aloud—literally howling—in my sleep. He later described it to me as a soft, plaintive yowl, distinctly like that of a frightened animal. I recalled that my cry, when I heard it inside my dream, had been loud and resonant. I wasn't scared, I was hurt. When Max woke me from my nightmare, I found that my entire abdomen—belly, back, pelvis—was locked in one immense, relentless cramp. My muscles throbbed as in concert, they clutched and clutched, never releasing between clutches. This soon became a regular occurrence: I'd come to in the dark with my own whimper bleeding through my consciousness into waking life and my body twisted into a feedback loop. When my body did this it overtook me with the weary vulnerability that was intrinsic to my chronic pain. But there was something also very fierce about the experience. I awoke a little bit wilder every time.

I started back toward the trailhead. The ripples in the earth's crust shrugged off their damage and remained unyielding. Epochs, glaciers, gulls passed. Trees stood, vulnerable and all-powerful. Granite and pines. Fire and ice. Down and up, down and up. Infinity pools of blue nestled between geologic miscellany. Beyond the horizon there was only sky, thin moon visible in the afternoon light, smoke rolling in from the west.

Sunset, Wednesday
Rock Creek CA
Cabin 6

Rocks, rocks, rocks. Stripes in rocks. Field of streams, stumps, piney flats, sand sand sand, river lake. Lake lake lake lake. I love thinking of us each alone walking on separate sides of the same mountain range. Today I felt relaxed even though I was tired. The sun shone. The rivers flowed. Water water everywhere even in the dry rock canyons. Water!

Potato chips for dinner, tomorrow I see you,

love.

· 4 ·

Steam

The people came to the mountains in search of escape. The people came to the woods and trails, lakes and rivers, reservoirs and campgrounds. They came to get some air, see a view—to touch the entity or collection of entities that they thought of as nature. The people came out of obligation, tradition, desire, or lack. They came to break the confines of pandemic lockdown, four walls, too many screens, a bad boss, tight budget, screaming child, sick mom. The people came aspiring to recreation or family time. They came because they had a standing date with the mountains every summer, or because they had managed to improvise three days off in a row. They were white people, mainly, but also brown and Black, and they had all kinds of bodies and minds with all sorts of capabilities and preferences. Some were wealthy and some were not, but all had the means to access a vehicle and a campsite or cabin. Some came from cities—Los Angeles, San Francisco, Modesto—and some from towns and suburbs, places with generic nature-inspired names like Golden Oaks or Laurel Vista. Their cars were large and small: red sedans, white trucks, blue trucks, black SUVs, hoods as tall as my head, fenders banged up, grills custom-made. Vehicles towed other vehicles—motorboats and campers, Jet Skis and bicycles—and towers of gear balanced on their rooftops. From back seats peered children and roommates and grandparents. The vehicles queued like commuters on the roads. On bumpers and windows, stickers were displayed; from a few antennae, banners flew. And here and there full-sized flags paraded from flagpoles jimmied into the corners of truck beds. The emblems bore words or images, statements of loyalty to a brand, a team, a candidate, or a place that the people perceived as their own. It was the Friday before Labor Day weekend in the United States of America, and the people were on vacation.

The heat had come before the vacationers had. It welled up heavy from the Central Valley and leaked into the upper elevations, spreading into forests and mountains, shimmering the air everywhere. In the Sierra, where in summer a backpacker had to be prepared for squalls above the tree line, it was hot. Too hot. Where the vacationers had come from, it was hotter: 109 degrees in Sacramento, 117 in the inland empire near Los Angeles, 102 back on the coast where I lived.

I was among the people, but I was going the wrong way, leaving the mountains and moving toward the heat. I'd retrieved Max from our prearranged meeting point in Yosemite and driven the cinematic descent down Tioga Pass back toward the Eastern Sierra, where we crashed at the cabin for a final night. Now we were headed home. The sky was fuzzy with pink overtones, a light haze of smoke. In my rearview mirror, Mono Lake salted silently. We planned to take a scenic route home, heading north to circumnavigate Yosemite from the backside before crossing west over the Sierra range on a road named Monitor Pass. From there we'd move through the foothills—gold country—until the highway split east-west, at which point we'd take a left at the Nevada border and go straight all the way home. Full tank. AC on. Good to go.

The road curved through and behind hills and onto a sort of high plateau, with pastures on either side of the road. Cattle grazed, seemingly oblivious to the exalted profile of the mountain range behind them. I navigated using the paper map that lived in our car, and Max drove. On the map, I traced our route with my finger. Out the window I saw creeks and shrubs, pine trees and cattle. I saw bustling small towns and towns with nothing but a dollar store. Then I saw the smoke cloud.

It billowed clumsily like a too-obvious punch line in the direction we were going: the smoke appeared to be coming from the road we planned to take, one of only two routes across the mountains in these parts. As our turnoff approached, so did the smoke.

We passed a rural school and noticed an armada of pickup trucks in the parking lot; a helicopter was parked on the football field. Fire trucks of various provenances were lined up in a rainbow on the black-top playground, and men checked equipment and hoses near a play structure. Rows of tents and shade shelters flapped in the dry wind, holding food and coffee and whiteboards with maps on them. I had never seen a wildfire command post before, and it looked like what it was: staging for a military operation. But if a wildfire was a battle between man and the elements, it was already clear which side would win the war.

Although I chafed at the sight of the command post, the military structure of firefighting made pragmatic sense; such large and quick mobilizations needed a hierarchy, and they indisputably saved lives. Despite the militant metaphors used to describe firefighting, it was more of an industry sector than an army, involving infinite layers of agencies, contractors, suppliers, and commerce. In 2020, emergency wildland fire suppression—*emergency* referring to wildland firefights that lasted longer than a day—cost Cal Fire more than $1 billion. At the time, the agency's total wildland fire suppression budget was $3 billion; that figure didn't include federal or local fire funding.

In practice, twenty-first-century wildland fire suppression was equal parts military strategy and manual labor, with a generous dash of theatrics. The main principle of firefighting that I had learned in childhood—water puts out fires—was an oversimplification. Wildfires weren't put out, anyway; they were fought, and sometimes contained. Containment wasn't extinguishment; the containment percentages I read in the news indicated a fire had been kept from spreading, not that it had stopped burning. Water was used, of course, but large-scale wildland fire suppression was often equally about dirt. One essential tactic was cutting line, which entailed people on foot, often in harsh terrain, using hand tools to clear every speck of vegetation from a strip of soil, thereby creating a fuel break, also called a firebreak or hand line, that deprived the fire of fuel and prevented it from spreading farther. The line might be as wide as a freeway or as narrow as a shovel, depending on the terrain and its accessibility. In larger-scale landscapes, bulldozers and feller bunchers, logging tractors that can grip whole trees in their claws, might be used to create larger firebreaks. These fire lines, also called dozer lines or CATlines, could help halt wildfires; they could also lead to soil erosion, water contamination, tree loss, and disruptions to wildlife patterns and sensitive ecologies. The heavier the equipment used, the more severe the damage to the land.

Additionally, fire was fought with fire: firing operations included backfires, in which new blazes were intentionally set ahead of or behind the main blaze, a tactic most commonly employed in areas with no or few structures. Ideally, a backfire was a carefully planned, measured operation. But in the middle of fast, intense, out-of-control blazes, they were sometimes used in haste or panic. And fire was fire: firing operations had the capability to get out of control or become unintentionally destructive, contributing to the size, severity, and lastingness of a wildfire.

All these tactics were carried out by legions of people, workers in an industry. Wildland firefighters, who were either employed by a spiderweb of various agencies or enlisted for as little as $2 a day from California's overflowing prisons, inarguably did essential work. They saved lives. In the public eye they were heroes, but in their own lives the recent increase in unfightable megafires was beginning to take a toll. The pay was low, the conditions dangerous, and the long-term health and mental health consequences of the job were becoming increasingly fatal. Worse perhaps, people who worked as firefighters while incarcerated were legally prevented from finding the same work once they were freed. In the United States, Black people, Indigenous people, and people of color were overwhelmingly subject to higher rates of incarceration. In late 2020, the California legislature passed a bill making it possible for formerly incarcerated firefighters to apply to have their records expunged after release, opening some paths to employment. That same year, the federal minimum wage for firefighters was raised—to a mere $15 an hour.

As we continued our approach to the pass and the hovering smoke cloud, I heard an airplane buzz overhead. In the public's perception, firefighters may have been the soldiers, but helicopters and airplanes were the big guns. The machines were where the theater happened. I'd seen footage of residents applauding as large tanker planes dropped harmful chemical fire suppressants over the landscape where they lived, or helicopters delivered their payload of giant baskets of water drawn from drought-depleted rivers and lakes. One informal term for such air attacks, derived from military machine-gun terminology, was *spray and pray.* A wildland firefighter once described to me a feedback cycle in which there developed an expectation among the general public that wildland firefighting meant large-scale machinery like tanker planes; accordingly, fire officials may feel pressured to use the big guns, and so the public perception would be reinforced, and so on.

As fire exclusion and associated megafires had become the norm, land management and firefighting became locked in many self-perpetuating cycles. In Northern California's Klamath River region, the Karuk Tribe had been working for decades to restore more traditional ways of existence; they found that current land management tactics under fire exclusion have severely impacted their cultural resources. During the early 2000s, sociologist Kari Marie Norgaard followed up her study of climate denial in Europe by immersing herself in the Karuk's organizing efforts, ultimately becoming an environmental justice policy advisor for the Tribe. In *Salmon and Acorns Feed Our People,* she documented how land management approaches that doubled down on fire exclusion decimated traditional aboriginal foodways such as salmon and acorns and destroyed important archaeological sites and artifacts—including what Norgaard described as the "cultural legacy of vegetation" itself. In addition, colonialist erasure of Indigenous worldviews through fire-related land management had further damaged "notions of belonging, responsibility, and reciprocity" regarding the natural world.

Past the school and fire command post, our turnoff approached. The smoke cloud grew. We had seen a few police cars near the command center, but there was no roadblock ahead, no indication of an authoritative eye on our location. We debated. Should we keep going, is it safe, what if the road is closed, what if it's not closed but it's not safe? Three-quarters of a mile. One thousand feet. Six hundred feet. As Max braked before the turnoff, a lone pickup truck descended Monitor Pass, coming from where we were headed. As far as I could tell, the man driving was not a firefighter. Well, if that dude made it through, we can make it, Max said, and we began the climb over the Sierra Crest.

A parched landscape beneath a brown smoke plume. Nobody lived on this road. With each switchback the smoke cloud appeared closer, but it was still impossible to site its location; it was so difficult to perceive depth in the mountains. Max fixed his gaze, steering, as the road became narrower, steeper. Another switchback, another sheer drop outside my window. Miles Davis blew cool sensations around the car on repeat. Time stretched, and we zigged higher until we were far above the world. I felt tired; Max and I had kept moving, as we'd planned on the beach in Santa Cruz a week ago, but fire followed us wherever we went.

Fire had come to this road before, but I couldn't tell how recently. According to a nature guide about the Sierra that I had read, we were supposed to be driving through glittering aspen trees and fluffy cottonwoods just beginning to show fall foliage. The trees we passed, however, were denuded and craggy, their skin the telltale black of prior burning. I did disaster math in my head, guesstimating the likelihood of a recently burned area igniting again. I knew from recent fires in Sonoma and Napa Counties that one burn didn't necessarily preclude another. One of the problems with the higher-intensity wildfires of recent years was that they led to reburning, sooner; when a destructive fire killed everything in an area, bushy and highly flammable invasive plants were often the first to grow back. Previous burning wasn't a guarantee of safety, as intentional fire wasn't a magic bullet but rather one component of an ongoing practice of care. Out my window I saw low, shrubby vegetation and not much else. Was the probability of this area reburning today low enough to make what Max and I were doing slightly less stupid? What we were doing: driving with spotty cell reception over an unpopulated mountain, some part of which was already on fire.

I steadied my phone and took a video: a long shot of the pyro-cumulus visible just past Max's shoulder, the easy, doped-up loops of Miles's trumpet lifting above it. At the summit, a sign read ELEVATION 8,314. The road swung left around the hip of the mountain and placed the smoke plume behind us. The cloud shrank as we moved west. We were through.

It would be only a year before fire would come back to Monitor Pass. When it came, the aspens would glitter again as charcoal, and the pavement would drip into rivers of melted tar around the lonely curves of the road.

I switched out Miles for Patsy Cline, still seeking familiarity in the past. Max drove on and down through foothill towns nestled between pine trees turned crispy by invasive beetles. Douglas firs teetered over the highway, felled by last winter's snow. These were ski destinations in other seasons, but in late summer's dry heat the landscape looked ungainly. As we descended, the recreational hordes kept coming at us—there was proper traffic now—from the cities and towns, seeking crystalline waters and fresh mountain air. They arrived to their weekend destinations tired and hot. They unhooked trailers, argued about tent poles, opened beers, and kicked back. TGIF.

In the Sierra National Forest, just before sunset that evening, another new wildfire stirred. Its incipient spark found abundant fuel, heat, and the strong, oxygenated wind of mountain valleys, and it reacted in turn. The Creek Fire surged through twenty thousand acres within the span of a day. In its early hours, at a crowded reservoir a few miles from where I'd hiked alone two days ago, three hundred vacationers were corralled toward the lake by flames on all sides. They gathered at the water's edge, trapped; the only road out was on fire. As their campsites and cars burned, the people waited in the lake, some of them ducking underwater to avoid the heat. The military apparatus swung into action. That night, two helicopters swerved through columns of smoke so heavy the pilots had to stifle the urge to vomit. They landed lakeside, tremulous ripples of water echoing the flames in the canopy, and the people were taken up. They wore shorts and flip-flops and baseball hats, and a few had surgical masks. They carried no luggage. Inside the windowless cargo bay the people sat pressed close, the collective weight of them testing the vehicle's thrust. The pilots found an opening in the smoke ceiling and took it. The next morning, all over the country, viewers watched footage of the rescue. Helicopter on tarmac, people ducking still-swirling blades, spotlighted in a violet light that emanated from the machine. On CNN, a rep from the County of Los Angeles Fire Department did PR. Despite the insurmountable conditions—triple-digit temperatures, multiple infernos, gusty winds—the young man appeared calm and confident, approaching cocky. This is something that we train for, he assured the camera. This is something that we prepare for, and we'll handle it.

Foothills. Gold country. Main street. Old West façades crowned low-slung, sunbaked buildings. From a storefront hung a three-quarter-sized effigy of a man, a noose around his neck. He didn't swing; there wasn't any wind. WELCOME TO HANGTOWN read the sign outside city limits, with PLACERVILLE in smaller letters fading into the painted wood.

The hanged man threw a grim shadow, summoning the specter of lynchings, a history that California shared with much of the nation. On the two-block main street, a shuttered antiques store located in an old movie theater had a marquee that read, THANKS COMMUNIST CHINA FOR TH CORONAVIRUS. These sights were a reminder of something I'd learned when I first moved to the woods: California was a wildly diverse place in every way, but despite this fact (and sometimes as a response to it), many parts of the state remained hostile to people of color, immigrants, and visibly queer folk. In the city I hadn't been rich enough to stay, but once I was displaced I found that rural California was available to me in a way it wasn't to many others. As a rural gentrifier, I didn't exactly blend in—I was politically to the left and made more money than a lot of rural residents—but I was white enough to feel safe in a majority-white locale. As a day tourist, a hanging effigy and displays of racist propaganda might make me uncomfortable, but I could walk past them and still experience this struggling American small town as rustic and cute. Max and I parked the car and used our white privilege to promenade down the old-timey main drag, in search of lunch.

In downtown Hangtown, the good sandwich place—Max had eaten there on a past work trip, organizing with casino workers nearby—had already closed for the day. We stopped to look at the lobby of the gold rush–era hotel, which was empty of employees and had a sign asking guests to call a phone number in order to check in. I stretched my car-sore hips while Max read display signs about the hotel's past as a way station for settlers. On the next block, the door to a small used bookstore was open. Inside, I found abundantly stocked shelves, whimsically named sections and genres, and a friendly middle-aged white lady at the counter. Max dove into the cavernous back aisles of the store to look for labor history books, and I made for the Californiana section. I found it easily, right up front, optimal placement for the many tourists stopping here on their way to and from the mountains. The section contained the same assortment of titles I'd seen in similar sections in countless tourist-town bookstores. There were two shelves of books about the gold rush, most of which used the phrase "Wild West" in their jacket copy. There were one or two outdated books about natural disasters, which was my preferred California beat—fire, earthquake, flood. The fire books were all published prior to 2010, and they all had red flames on the covers; the earthquake ones all featured photos of collapsed Victorian buildings in San Francisco. A few photo-packed titles looked back with fond nostalgia on the hippie generation. A slim selection of Hollywood-centric nonfiction focused on celebrity memoirs and true crime tales about murdered starlets. The California History subsection consisted entirely of books about the missions written from the perspective of the missionaries. A few books focused on Native American culture, but the jackets cringingly deployed tropes of extinct, magical "Indians" without acknowledging the continued survival and diverse interests of Indigenous people; the authors appeared to be white. These literatures of California told the usual lies of cowboys and Indians, wildness and new frontiers. I was sick of them all. What did I know about this state that was actually true, I thought as I browsed.

California was a place of rapturous beauty, a land that for millennia had inspired wonder at the magnitude and majesty of the natural systems on display there. And California was a place predicated on the manipulation of its natural systems and resources, sometimes out of care and sometimes out of harm. It was a place of violence: the violence of the land; the acts committed by people seeking to claim it as our own; and the violent systems people had harnessed to enable human dominance on the landscape. There were so many stories about this land—the mountains, the forests, the cities they'd built—but many more had been hidden. Just as a person might piece together an origin story that reflected only one perspective on a multifaceted life, people created stories about their experiences with places that grew into assumed truths. To its thirty-nine million residents, California meant thirty-nine million things, but from the looks of that bookshelf it appeared only a handful of them had made it into wide circulation. My own California story had stemmed from my parents' pursuit of a uniquely Californian myth—the hippie movement's search for new horizons, real or imagined. To construct my California had taken a lifetime of choices, elevating some aspects of the place where I lived above others. The only thing about the West that I knew to be a fixed truth was that this was where the sun touched the land last, every day. We were always meeting the light head-on, even when it burned too hot.

We stopped to fill up at a gas station by the freeway on-ramp. Vacationers filled the tanks of vehicles towing other vehicles. On a shiny trailered speedboat, four full-sized flags hung limp in the heat. One bore a photoshopped image of the fascist president, shirtless and tan. One looked like a regular American flag, but rendered in black and white. The bars on the flag were designed to appear shredded by claw marks, and instead of the usual stars, there was a cryptic set of numbers. Another flag showed two crossed rifles cradled in the talons of an eagle, and the final standard displayed a snake against a yellow background. It was the symbology of fascism and white nationalism. In the truck's window I saw a sticker of a handgun barrel pointed at me. As I squeegeed the windows and Max pumped the gas, I again experienced the sensation of slippage: a feeling of coming ineluctably closer to free fall. The soap from the squeegee met the dirt on the car and dripped long lines of mud toward the ground. Max and I exchanged a silent look, well practiced over more than a decade of traveling together: Let's get out of this place, the look said.

West from Placerville the traffic eased. Evening approached. A glutinous mass of smoke filled the Central Valley, locked in place by the mountain ranges on either side. The smoke lingered low in farm fields already dusted with chemicals and fertilizers. In a large block of fruit trees, people in hooded sweatshirts tended the crops. Almonds, wine. Water, gold. Culture, sex, nostalgia, tech. We took and took, and California kept giving.

Ten years earlier, in what my stepmom liked to call the days of young love, Max and I had traveled around Europe on bicycles for six months. We had both been new to long-distance bike touring, and as soon as I discovered I could tough-cookie myself over the Pyrenees carrying everything I needed strapped to my bike, the landscapes through which I traveled became transformed. I pushed my bike and body across endless rolling farmlands in southwestern France; deep evergreen forest valleys in northern Portugal; dry oak chaparral in central Catalonia; and small Sicilian towns encircled by walls from ancient wars. On a bicycle, roads were rarely flat; what looked like easy, rolling hills in a car could be an endless sweaty afternoon on a bike. I spent many long hours looking at the ground as it rolled beneath me, observing the landscape closely. There was something about traveling in such a fashion that slowed time, made nothing urgent. When I crossed a landscape slowly, at the level of the land, I discovered its contours in a direct way—by touching it. After six months living on the bike, I could look at a road and understand where it might be steep and where it would curve, or how a footpath I assumed was a straight shot through a grassy field might end up plunging me into sheepy downs in impenetrable mist, only to turn corner after corner, twisting, ultimately allowing me a way forward only by winding me back around the mountain again. Every aspect of the ground beneath me was palpable, registering in my body across the full spectrum of sensation.

Max and I drove on toward the Sacramento Delta, and the light sank lower behind a veil of particulate matter. Max fiddled with the radio, flirting with new country and rock en español. I wiggled in the passenger seat as comfort evaded my body. Even my pain was road-weary. I propped my bare feet up on the dash, and I tried to recall that bicycler's way of sensing the land. First, I noted the shape of the earth beneath the highway, each time the road cut or bent in a way contrary to the flow of the terrain. Then I looked at the hills, peach-fuzzy as human skin where they met in the bosoms of plump valleys. I saw the way a riverbed always finds low ground, no matter where the houses are. From the highway I tried to picture those great, long-dry rivers bigger than anything we had now: the Merced River when it was still mighty, pouring down from the High Sierra, shaping this valley. Mountains deep as water; earth black as rock; air like a sprite that can never be caught.

The valley turned into the deep East Bay, where highways cut cross sections out of the landscape. Subdivision after subdivision, some still under construction, grew where there once must have been grasslands and chaparral. If I squinted my eyes I could skip over these blocky objects, the mark of human presence. I pictured the land as empty, these settlements as a stain. I tried to envision how wildfire would have coursed through here in earlier eras, the routes it would have taken across differing fuels. Fire or water or animals might have sped along the cut of the highway I was now on, or at other times run crosswise against its flow. Earlier eras may have been cooler than this one, but still the warmth of a second-summer afternoon on dry grass and oak would have caught fire easily. As could the houses that now covered these hills.

No one should be living here, I said to Max, meaning, in this place so obviously dry and fire-prone.

But everyone lives somewhere, he said.

Can't you just picture it without us, though, without all this crap, I said.

The ideal of a pristine wilderness was still so tempting. It had grabbed on to some deep part of me—perhaps something in my whiteness or my inherited aesthetics, the latent settler inside me—with a familiar touch. Even after I had learned to notice it, I found it difficult to untrain myself from the romance of an unpeopled California.

In the young days of middle-aged love, when we bought the house in the woods, we still for a time had my old rent-controlled apartment in San Francisco. There was a transitional era in which we lived in both places: the redwoods and the Mission District. Each Friday I loaded into our hatchback a huge blue plastic shopping bag from a chain store, containing our laundry, and an insulated food bag from a different chain store, containing the perishables from the fridge, and drove two hours through traffic to the little white house under big red trees. (Max usually commuted separately, as he worked long and odd hours.) At first, there was whiplash whenever I got out of the car in one place or another: time moved differently in the city and the country. But as my life adjusted to its new rhythms, the contrast between the two environments began to diminish. On my way from the house in the woods to the apartment in the city, heavy tree cover segued into the open cow pastures of southern Sonoma County, which then gently unfolded into the suburban density of Marin County, Mount Tam always a surprise behind the scrum of strip malls and multimillion-dollar homes, until the towers of the Golden Gate Bridge segued easily into the urban environment. On the way north, skyline turned into fog; the fog enfolded the bridge; the bridge's orange span then bled again into the golden ombré of the Marin Headlands. In this way, instead of understanding a city as an anomaly in the landscape, a thing apart from the green and open things around it, I had begun to see how a peopled place might simply be another type of Earth's architecture. The built environment wasn't as dissonant to the natural environment as I had assumed; in fact, they were so clearly the same thing. Concrete canyons had discrete wind patterns; streets were paved over the paths of now-subterranean creeks. Even the pollution emanating from the Chevron refinery, whose caches and coffers loomed, villain-esque, over the bayside in Richmond, was part of the complex whole.

In the car on Interstate 80, I pushed back against the allure of an empty wilderness and tried to de-settlerize my view. Instead of picturing the landscape unpeopled, what might change if I pictured the people as part of the land?

Earth, water, air, fire. All things came from these elements, in some way. The cul-de-sacs we passed were built with wood and stone. The road was made of petroleum. And what was a highway sign, anyway? Metal, paint, screws—objects and materials derived from parts of the planet and shaped by people using the elements as tools. Fire shaped metal. Fire made heat, which made chemicals, which made paint. Fire was what made infrastructure and weapons, plastic and leather and cars, cars, cars surrounding me and beneath me, each one a smoking gun. Inside the cars, gasoline—this too came from fire, fossils of dead trees and critters, dredged up or blasted or dug and transformed by fire first into steam, then coal, fuel, oil—that black gold for which humans were willing to sacrifice ourselves. Fire was inside the origin of everything: a molten core. Fire was central to the transformation of Earth's material into fuel and property, and to the uncontrollable effects of all that alchemy on the planet's ecosystems; fire was man's perceived power and man's great fear. The history of people was a pyronatural one, as was that of capital and empire. I wondered if Prometheus, when he stole fire from the sky for us, knew what we'd do with it.

As it got dark we drove past a new burn scar near the city of Vacaville, which had once been a remote farm town until the Bay Area's growth overtook it. A few nights earlier the LNU Lightning Complex—the same fire complex that was also burning near our home, an hour north and west of here—had poured over the eight-lane highway and inundated the tract houses and ranches at the edge of the WUI. Fire could make the materials of human life, and it could unmake them too. The burned air scratched my body from the inside. The countryside out here used to be grass and oaks. Now it was black dirt. The hills were nude. Fire had laid bare the contours of the landscape until there was nothing but the land itself: nothing left, everything possible.

September

· 5 ·

Hawk

In 1778 a French gentleman farmer, who had settled in the region of the Algonquian-speaking people in the place newly called Pennsylvania, wrote to a friend back home. His dispatches described an autumnal period in which the weather cooled and was then followed by "a short interval of smoke and mildness, called the Indian Summer." It wasn't known whether the smoke he mentioned was literal, as early fall was a period of time in which Indigenous people in that area used fire to steward the land. Perhaps he was merely describing the halcyon feeling of the brief weeks when summer seems unwilling to let go its grasp on the land. Regardless, *Indian summer* traveled back to Europe around the same time and wound its way into European culture, then back again to the Americas, where it stuck, right next to slurs like *Indian givers*—language that equated indigeneity with something that was false. Until 2020, "Indian summer" was the American Meteorological Society's official definition of a period of warm weather in autumn.

In California it had been hot in autumn my entire life. Growing up I had referred to the season as Indian summer without thinking to ask why it was named that. In my teen years I came to understand the term as one to avoid; it was derogatory, I knew, but that was as far as my inquiry went. By the time I was an adult I was instead saying second summer—the weeks or months when the marine layer that cooled the coast wavered and disappeared, and the weather became dry and hot. It was relatively recently, long after the time of childhood mythologies but sometime before my first wildfire evacuation, that I found myself using a new name for autumn: fire season.

Fire season was a loosely defined term invoked by Cal Fire, local authorities, and the media to describe the dry months in the American West when fuels were dry, the weather was hot, and wind drove ignition of wildfires. In the greater Bay Area, fire season was primarily in September and October. This year we had already been through enough fires for several seasons, and autumn was only beginning.

Beyond its literality the term fire season seemed overnight to have become a sort of catchall to describe not just the weather but the state of the world in general. Fire season had once been a semi-technical term, but now the vernacular upped the ante. The phrase seemed to encapsulate the experience of living in a moment that felt precipitously like a decline—that slippage again. Fire season could mean anything that was unpredictable and uncontrollable, which at times that autumn felt like everything. If a person made plans to meet up with a friend, there was always a caveat: But you never know what's going to happen, it's fire season! When the grocery store clerk asked how you are? Well, it's fire season! That creeping feeling that bad things were going to continue to happen? It must be fire season. A neighbor stopped her car by my driveway to ask if the monthly fire-safety meeting would be outdoors, for social distancing. Who knew?! I shrugged. It was fire season! Fire season could be a response to any question, a framing for any narrative. For those of us living through it, the idea of fire being a season was also an expression of hope, or perhaps wishful thinking: if fire was a season, that meant it was temporary, and at some point it would go away.

By the time Max and I returned to the house on the hill in the woods, our zone's evacuation warning had been rescinded. The decks were dirty with ash and fallen redwood duff. Inside, the house smelled like smoke, but it was still cooler than outside. There was an uneasy quiet to the forest. In the living room, an abandoned bouquet of sunflowers had dropped its petals on the coffee table. The milk had gone bad in the fridge. The clock on the oven blinked, indicating that the power had been out at some point. We dumped our camping stuff by the door and went to bed. In the morning Max made breakfast, we did laundry, and I checked my maps. The fires were still burning, the season relentless. In Cazadero, a town ten miles northwest of us, residents with bulldozers were scraping a six-foot-wide line in the dirt to keep the Walbridge away from their homes. The smoke had stayed put, suffocating the city folk too. Fire wasn't going away. Fire was everywhere, it would continue to be everywhere, there was nowhere it couldn't reach. It was inside me right now, I thought, as I took breath after breath. I put on an N95 mask and went outside to find something pretty to look at.

In the garden it was 106 degrees Fahrenheit. According to my air quality app, it was dangerous to be outside: the air quality index was 250, which Max and I called purple to correspond with its color on the official AQI scale. Purple was second worst. The language that accompanied it on the charts read, "The risk of health effects is increased for everyone." Purple was topped only by dark purple, sometimes labeled maroon—"Emergency conditions: everyone is more likely to be affected." Beyond an ever-present thud inside my temples and bottomless fatigue, the charts were unclear about what those effects were. Medical and science fields were only beginning to understand the long-term health effects of wildfire smoke, but when the chart got into redder hues the takeaway was clear: don't breathe the air.

The yard was sepia toned and slightly out of focus. Weeks of heat and smoke had turned the flowers and trees into memories. I was outside to check on the garden and hunt for lung herbs, plants with medicinal qualities that, when brewed into a tea, might help my body cope with smoke. Social media was full of recipes for throat soothers and firefighter tea but I already knew which herbs to look for. Growing up in a hippie town had its perks. I had in my sights mint, rosemary, mullein, and lemon balm.

Rosemary grew in the wine-barrel herb garden just outside the back door. Mint ran riot in a flower bed (never again, I reminded myself, as it spread and spread). I knew there was a volunteer mullein plant in the less-tended part of the yard down the hill, so I did a slow lap on my way: over browned grass, past the Meyer lemon tree, past the flower beds, through the jasmine arch, and down the slope past the orchard. Fallen branches and twigs were all over, scattered remnants of the wild wind. I felt the soft brush of native hazelnut bushes against my shoulders. Lemon balm clustered in my wake. In the shade beneath a family of redwoods I found violets growing, and they gave me their heart-shaped leaves. Returning up the path toward the irrigated part of the garden, my lungs worked hard on the incline.

Next, I patrolled the flower beds. I greeted each plant, not verbally but by directing my attention to it. I was shy around the apple trees after our last farewell. It was that time in the growing season in California when the garden was awash in reds and oranges. Yellow sunflowers taller than I was leaned on stalks of red amaranth, which drooped under their own weight. Closer to the ground, yellow petals and black eyes of rudbeckia sprawled in gawky groups. In the heat most of the flowers were angled downward, shying away from the sky. But among apricot hues of roses and coral clusters of yarrow, a coterie of purple coneflowers stood tall: *Echinacea purpurea,* its petals thrust backward in an open, reverse-daisy formation that to me appeared always up for anything. I grinned at them. The dahlias were between rounds of flowering, the lowest tiers of their foliage starting to turn from green to a pale yellow. Traces of something white were splattered like paint on a few leaves; I couldn't tell if it was powdery mildew, an undesirable garden fungus, or ash. I put my herbs in the pocket of my sweatpants, where I'd forget them and find them later. In the center of a pink zinnia, more ash and bits of burned matter nestled amid the flower's stigma and anthers. This was the time of year when most mornings I found bumblebees asleep inside flowers, drunk on pollen and hard work. I wondered what wildfire smoke did to bees. My head felt pressurized. Inside my pelvis, my nerves felt singed.

With my pruning shears, a housewarming present from my dad, I started instinctively to deadhead, cutting off spent blooms and tossing the clippings into the path. We'd been watering less frequently because of the drought and I could tell the roses were less vigorous than they would be otherwise. Some of the roses were burned from the heat. The edges of their petals were crispy and darkened. A few were about to open, but I knew the sun would fry them as soon as they did. I hated to see them diminish in the heat. I kept deadheading. As I moved through the beds I gathered the beginnings of a bouquet. I couldn't save them all from the weather, but I could invite some of them inside and partake of their beauty in the meantime.

I was wearing an old button-up shirt in a loud flower print, collar and cuffs closed tight against the sun. My baseball hat had a flower on it, too. On my arms bloomed tattoos of roses, vetch, and trillium. My shirt snagged on a yellow rosebush, named for chef Julia Child because it looked like butter. I had to shift backward at the same angle in order to free myself from the thorn, and as I moved a different cane scratched my forearm. At the bush's top, about even with my head, there was a stem bearing three rosebuds, each about to open. The sepals around the buds' bases were open in anticipation but the blooms were still closed. I squeezed one and it gave like a marshmallow. Perfect. Cut.

Before I moved to the house in the woods, I never had a garden. The house came with two abandoned rosebushes amid the bramble; we dug beds around them and gave them neighbors. More roses, bulbs, salvias—all flowering, I wanted only flowers, flowers and fruit. Beauty and sweetness. Early in the pandemic we had tried to grow the odd radish patch and climbing bean, but soon relapsed into flowers. I often joked that, compared to friends' gardens overflowing with summer squash and snap peas, all my little flowers were useless. But I grew them because they were the only things that made me feel good. Beauty wasn't merely an aesthetic stipulation of my garden; it was medicine.

After the last surgeon removed my reproductive organs, the pain I'd experienced throughout my initial medical accidents and incidents diminished enough for me to accomplish daily tasks and activities. For example, I could walk now. But still those injured, animal sensations were with me more days than not. The doctors couldn't say what specifically caused my pain to become chronic, but in my experience doctors tended to be caught up in overly restrictive definitions of cause and effect. I had an abundance of inciting incidents to choose from. Pain was a poorly understood medical phenomenon—a symptom that was treated as a condition when it became chronic. My chronic pain was, generally, a relationship between nerve and muscle, a complex feedback loop of tangible sensation, neurotransmitters, and lived experience. Some days the pain was a distant whine, one I barely noticed. Some days it was stabbing, the adjective well-earned, ripping through the interlaced levator ani muscle group in my pelvic floor. The pain had many forms and moods. I learned them all. Dull ache, sharp spike, flutter. Pain twined up my neck; it sent feelers down into my hips and thighs. My physical therapist theorized that my muscles had become permanently stuck in the act of clenching, defending my core against assaults that were long over, until the defense had become an assault of its own.

When I was in pain, which was some part of most days, some days more than others, it often felt to me as though there was nowhere to go. The pain burrowed inside every aspect of my being. It gave my body and brain no exit. The locus of each spasm was rarely pinpointable—it often felt like a tiny Zeus held court inside the void where my uterus used to be, assaulting me with his little lightning bolts, or sometimes it simply felt as though I'd done too many sit-ups. I compared the relentlessness of chronic pain to a fishhook, tugging me downward into a deep whirlpool of distraction, a perpetual undertow. Pain had a way of sneaking up on me. When I used to smoke cigarettes, I enjoyed lighting them with matches. Whenever I struck a match and was slow in bringing it to my mouth, the matchstick would begin to burn up before reaching my finger and thumb, but I wouldn't know it until at once, rapidly, I felt the skin of my fingers burning. That moment, the burn point: that was pain. And, like fire, pain seemed alive. It devoured and traveled. It evaded description. It was formless yet ever present. Sometimes it just felt like sadness.

Flowers were the only thing that felt good. Like most plants, flowers could be medicine or poison. Since the pandemic had begun there had been an increase in publicity about the restorative power of the outdoors. I had clicked on studies and articles and read entire books about the healing properties of plants and gardening, forest bathing (which, as it turned out, was just hiking), and plant medicine. But none of the literature shed light on what I already instinctively knew:

The main reason flowers made me feel good was that they were beautiful. They were joy. They also weren't exactly useless; flowers were key to increasing biodiversity in an ecosystem. Pollinators used them for food and shelter—and without pollinators there would be no food.

A towhee chirped from a bush somewhere. A raven snarked back at it from on high, asserting dominance. A bumblebee buzzed my head, not drunk yet. I took off my mask, telling myself the air quality wasn't that bad. The garden smelled like ash, but also roses—almost artificial in their sweetness—and also a bit like a skunk, which I realized was because I was standing downwind from one of three cannabis plants interspersed in the flower beds, for personal use, as the law said. The redwoods let off an organic scent that came not from the canopy above but from the ground below: fallen needles in layer after layer formed the soil of this place. I sat on a wooden lawn chair with my handful of blooms, and I asked my body to relax.

To help, I began to build my bouquet. For the roses, I used pruning shears to snip the thorns off. For the others, I held the stems upside down, so the buds wouldn't open too widely yet, and ran my hand up each stem, removing the leaves with the speed of the gesture, then tossing them aside. They'd compost where they fell, eventually. I made my left hand into a sort of claw shape and began to arrange stems vertically between my fingers, my grip holding the arrangement in place. My sister Caroline taught me this technique, which some florists use to shape a bouquet instead of the noxious green foam that comes in many arrangements. I tilted and fussed with the stems until the blooms were pleasingly arrayed. To me, making a bouquet was similar to writing or editing. It was a creative act. It required attention and love and a knack for finding the exact odd angle at which to approach a kernel of beauty. When it was good, it was a unique combination of intentionality and instinct. When it was hard, it was a reminder that I was not in control.

The physical aspect of gardening often caused me pain—weeding especially, with so much bending and reaching—but it still felt restorative. I could be angry or self-pitying, succumbing to the undertow, but if I went outside and picked a little bouquet I became immediately calmer. At once, I could breathe again. I remembered that the sky was there, and that I was small in this big violent world, and in the garden somehow that knowledge made me feel better, not worse.

It took approximately four minutes of my being among flowers before my body began to respond. My breathing steadied. The grip of my pain loosened slightly. My pulse slowed, time slowed, and the garden smiled up at the redwoods. Every bouquet, every new planting and crisp winter morning spent weeding: as our relationship grew over time, the flowers that I grew for beauty's sake required more of my attention, more of my touch, linking my body more closely to the body of this place.

While I had said many times that my garden healed me, it was difficult for me to explain how, exactly, my body and the garden and the big violent world were linked. Yet they were. Why wouldn't they be? After my medical calamities began, friends and colleagues had suggested I write about what happened to me—as though an essay were a form of revenge. I soon gave up on writing about what happened because I couldn't find a way to talk about what happened after, in the garden. I couldn't summon language with which to describe the processes of nature—the magic, the metaphor, the feeling that there was some larger meaning, a great pattern of life that I fit into, somehow. Of course plants healed people, that was the way the systems of the natural world functioned. It was almost embarrassing, the mundanity of healing through contact with nature. And yet. There was that renewal. The young good green of early vegetative growth: proof of life.

To survive the experience of my body, I needed beauty, and I found it in flowers. But the flower garden was situated within an ecosystem that was itself trying to survive crisis. Every summer was hotter than the last, every winter harsher. It never rained anymore. There were weird insect invasions. The trees dropped dry duff earlier each fall. The tan oaks in the woods grew sicker. The wildfires came closer. This hillside was the place where the contradictions of my condition, my existence, pain and beauty, bad fire and good fire, the whole burning planet, were able to coexist. The messy diametrics of my body and the ecosystems it touched were to me evidence of connection, not an argument against it. Through my physical interaction with the land, via gardening, I was experiencing the ways in which I was a part of my ecosystem, not just an appreciator of its beauty. I felt that connection most directly when my body made labor in the garden impossible. On those days I found my way back to the chair that faced the sun in winter. I sat and let my body shake as the pain wrangled me into its void. I rested my head against the chair's back and caught sight of the redwoods dancing in the wind above, and I felt the same heat, the same light, spreading across our surfaces.

Satisfied with my improvised bouquet, I needed to take it inside and get it in water. I stepped past an apple tree to reach for one last addition to my posy—a cottony tuft of *Daucus carota,* also called Queen Anne's lace or wild carrot flower—and my hip roughly nudged a lateral branch on one of the young apple trees. It snapped and broke off. We were each of us damaged, perhaps that was enough.

If my growing obsession with the garden was about connection and beauty, it was also about bodily control and vulnerability—a reckoning with mortality. Gardens meant different things to different gardeners. The early-twentieth-century Czech playwright and author Karel Čapek, who coined the term *robot* in his science fiction works, wrote of his gardening hobby as a comical battle, a farce of man's cluelessness in which his primary antagonist was a tangled garden hose. The writer Jamaica Kincaid viewed gardening as a leisure activity but also as a sort of reclamation; Kincaid grew up in Antigua, a place where the activities of British colonization had once denuded the island of trees, and her garden in the American Northeast had evolved in part as "a conversation" with that legacy. To British aristocrat and writer Vita Sackville-West, horticulture was pure poetics, a language of beauty. For Sackville-West's lover, the author and publisher Virginia Woolf, gardens presented a spectacle of sanctuary that she could enjoy from her frequent sickbed. (Judging by Woolf's journals, it appeared that her husband and copublisher, Leonard, did most of the hands-on labor in their gardens.)

In my perpetual and unsuccessful search for health crisis stories that were corollaries to mine, I had recently read Woolf's 1926 essay "On Being Ill," an odd piece of literary criticism concerned with the absence of portrayals of illness in literature. Woolf herself suffered from depression, long-term effects of the 1918 flu, and other untold physical and mental maladies. She wrote from experience. In the essay, she observed that people who were ill often turned to nature in their time of invalidity. Illness, Woolf said, was an ideal state from which to regard the human condition, because ill people were outside of the daily grind; a sick person was exempted from onerous terrestrial tasks. (Woolf didn't acknowledge that some ill people might still have to go to work but, well, she had money.) It was from the unique vantage point of enforced rest, she said, that a person might truly appreciate the natural world. Whereas quotidian life was filled with responsibilities and tasks, an ailing person was permitted to recline, to stare out the window and do nothing else—to watch, for example, a rosebush grow slowly for an entire afternoon. "It is only the recumbent," Woolf wrote, "who knows what, after all, Nature is at no pains to conceal—that she in the end will conquer." She being Nature, not the recumbent. Let the next ice age come, challenged Woolf; we'd all be long dead, but somewhere, in the ruins of a garden, the flowers we planted would still bloom. In Woolf's scenario, humans looked to nature to heal because of nature's indifference, not in spite of it. The cycles of the natural world enforced upon people the eventuality of a sweet abdication of organic matter, the promise of a cycle that might someday conclude. Plants offered salve to the suffering not merely because they symbolized life but because they reminded us of its inevitable end.

Back inside, I threw my N95 on the table with the other masks. The left lower quadrant of my belly undulated, pain returning to my span of awareness—probably about a five out of ten on the pain scale today, I self-diagnosed. Not too bad. I placed one hand on my abdomen and one hand on my chest, as my pelvic floor physical therapist had taught me, then tried to breathe into and through the belly hand. The therapist said that when I did this deep breathing, I should try to imagine my pelvic floor opening, the way a flower opened, one petal at a time. Like those nature films that show a time-lapse sequence of a rose in bloom, she had said. I breathed the stale indoor air and tried to picture my pelvis unfurling like a rosebud. It didn't feel like a flower; it felt like a bear trap.

After we de-evacuated it was difficult to care for my home or my life. Upon my return from the High Sierra, I did not want to clean or cook or fix the crooked ceiling fan or even garden, much, certainly not hike, although the smoke kept me indoors anyway. A sort of subterranean reluctance, which had been with me since Santa Cruz, weighed me down. There was a monotony to living in a constant state of alarm; I moved slowly. I didn't unpack the bags in case we had to leave again.

By Monday—Labor Day, the unofficial last day of summer—all the trapped vacationers had been evacuated from the new Creek Fire in Sierra National Forest, and every national forest in the state had been closed as a precaution; within two weeks, Yosemite too would close due to heavy smoke. In the North Bay our fires reached containment in the double digits. In Sonoma County evacuations ended, but inland, in Butte County and Mendocino and the Sierra, the daily rollout of damage statistics and evac zone maps plodded on. At the little white house in the former logging camp the air quality improved from very very bad to just plain bad. In the morning, Max and I attended our neighborhood fire-safety meeting—about a dozen white people in masks on lawn chairs in a vacant lot beneath a ring of redwood trees. A handful of mottled dogs napped in the middle of the circle. My neighbors were talkative but not particularly productive. On the agenda: an evacuation phone tree; a proposal to organize a fuel reduction workday on the hill behind our house; a reminder to look up home-hardening tips on the Cal Fire website. I listened to the agenda items and mentally rolled my eyes at each one. Now that we lived inside nonstop disaster, preparing for it felt hollow to me. It doesn't matter, I thought. It's already happened. It is happening.

The heat was unendurable. The dogs sighed in their sleep on the dirt.

After the meeting I sheltered in the relative cool of the house, windows closed, ignoring the recurrent feeling of suffocation in my chest. Some friends, Jesse and Dani, texted: They were headed to one of the less-busy beaches on the river, the AQI wasn't too bad today, did we want to come?

The river was a green thread. Large trees stood above their reflections in the water: redwood, fir, buckeye. I had heard from an old local fisherman that only a couple of decades ago salmon ran in bountiful numbers in this river. Now scientists and ecological activists were struggling to revive the population. Across the shore was a row of vacation rentals; empty wooden docks led to stairs.

Pebble beaches bit our feet. My friends and I walked into the water, hopping and tripping on the rocky riverbed. I attempted a brief backstroke but it felt too exerting. I threw off my sun hat and dunked my whole head and body. Underwater, it was cool.

Whenever I was in pain, a reliable way to abate that pain, however temporarily, was for my body to be in water. Hot was best, but any water was relief. In addition to frequent baths I had recently taken up swimming. Although I'd grown up playing in the Pacific, I hadn't swum in decades. The currents off the Sonoma Coast were often dangerously strong, and when I went to the ocean I stayed on the cliffs or the sand. I was also allergic to chlorine, which kept me out of pools. Earlier that summer my friend Susan had invited me to swim in the chlorine-free pool at her house as often as I wanted. I started small, once a week. I soon discovered that I had an aversion to putting my head underwater. I had been a strong swimmer as a child but after spending my twenties being cool and smoking cigarettes, I had developed asthma. Now every time I submerged my head I found that I couldn't breathe. My lungs contracted, which led to the instinct to inhale, which led to dangerous situations. This type of shock was a normal physical response to cold water, one I'd experienced before when jumping into glacial lakes or the Monterey Bay, but now it happened to me any time I tried to swim, even in warm water. In Susan's pool, under drooping oak trees beside her flower garden, I began trying to reteach myself to breathe correctly. To practice, I did separate parts of a crawl stroke: first I held the side of the pool and kicked with my legs while turning my head left to right and breathing in rhythm. Then I did the arms, walking through the shallow end while stroking and breathing, dipping my head in and out of the water. I made slow progress. At the end of each session I did a few laps of backstroke, relaxed, lying supine with the sun in my eyes, and I felt my lungs open to the air and my body move effortlessly, the way it wanted to.

The river was so shallow that I had to sit down to get fully underwater. I kept my eyes closed against algae and mud. As soon as my head was underwater, my lungs did their reopening thing, mistaking water for air. I surfaced coughing, the sky still made of smoke. The Walbridge Fire, which we had all just evacuated from, was about three miles north of where we swam. Another fire burned on the coast, three miles west.

Dani had a stand-up paddleboard and she and I took turns propelling each other around on it, one of us sitting on the board and the other walking in the water and pushing. Dani had grown up just over the hill from me in Palo Alto; she was a chill Latinx woman who had a demanding job in philanthropy and a knack for herbalism. Her partner, Jesse, was a white academic working on climate and labor issues; he had grown up directly across the river from where we now swam, and his mom still lived there. Theirs was an old activist West County family, outgoing and musical, and their lives seemed to me to be grounded in a sense of care for the world. Moving through the shallows, half walking and half swimming, I felt held by the water. Compared to the crowded air outside, the slow current felt clean: a breath.

Standing in the river, my friends and I caught up; I learned I wasn't the only person feeling numbed. Everyone had had a stressful evacuation (was there any other kind?). Everyone was depressed—everyone on Earth seemed depressed that year—and we were all stir-crazy from the smoke. Prior to the pandemic, Bay Area residents had experienced hazardous smoke from previous years' wildfires, and we knew the drill: stay inside, close the windows. After the necessary social restrictions of the pandemic had made gathering indoors a health hazard, the outdoors had become a respite, a chance to breathe freely and socialize safely—until the lightning fires enclosed us all in smoke again. Now everywhere was dangerous, inside and out.

The circumstances reminded Jesse of how he had felt at the start of the pandemic. He and Dani lived in Berkeley, but they had spent the duration of the shelter-in-place order at Jesse's mom's house here on the river. Some nights, Jesse said, he would get so claustrophobic he just couldn't handle it. This was when people had been afraid to leave the house at all. During those cloistered months, whenever Jesse experienced that trapped feeling, he would put on headphones and walk from his mom's house to the cliff above the riverbank. There, a small spot of concrete marked where the road used to be, until some long-ago flood forced it farther inshore. Jesse stood on the concrete patch and danced in the dark, volume up, in outward silence, shaking his limbs at the stars. He felt, he told us, like he had when he was growing up in this area, a river rat without an escape plan, overwhelmed by that universal teenage urge to *get outta here*. Now we might really have to get out of here, someone said. We said this all the time. The question of *how long will climate change let us live our lives this way* was by that time in my peer circles a well-worn ritual, a form of conversational solidarity that somehow neither comforted nor effected action.

I hadn't been to the river in a while, and I found I'd missed the ritual physicalities of beachgoing: stumbling out of the water over rocks, inserting muddy feet into shoes left at the water's edge, collapsing into the comfort of a sand-warmed towel. Greasy chips, lukewarm beers, a big hat. It felt like summer, teenage summer, the good kind. Despite the poor air quality and the knowledge of what was happening in the woods on either side of us, it was good to see friends. We were refreshed by one another's company; Jesse later described it as a desperate joy we made together on the river that day.

After we'd been at the beach awhile, rotating in and out of the water, I looked up from my towel to see Max talking to a middle-aged woman at the river's edge. Her feet were bare, she wore a hoodie, and her head was upturned to the sky. She was alone. She was white, with brown hair, and she looked tired. She could have been one of our mothers. She kept looking up to the mudded sky, then around at other beachgoers with an air that indicated neediness, like she really wanted to talk to someone. I often steered clear of people like that—I could swear they were drawn to me—but Max in his infinite extrovertedness liked to engage everyone. He greeted the woman and her story cascaded out: her house had just burned down.

In the Walbridge Fire? Max asked. Yes, she confirmed. The house had burned last week, but she had just seen it for the first time after sneaking past the police line in order to water her garden. She had fruit trees, she explained, they'd been burned, but not too badly; she didn't know if they'd survive. The house was gone. Cats, too, or at least they hadn't come back yet. Her kids had told her to stay away, but she had to go back and see it. To make sure it was real.

The woman said she lived up behind Armstrong Woods, a popular state park at the north edge of the river town of Guerneville. I knew it well. It was like a Platonic version of what the redwood forest where I lived might have been, had it not been logged then left unmanaged. Armstrong Woods was a small redwood valley; flat, wide paths led visitors on short hikes through groves of giants. Trailside signs detailed facts about native flora and provided micro-histories of the park's founders, a white logging family whom the sign hailed as "pioneers of conservation." A logger saving trees might strike some as unlikely, but during the lumber booms of the gold rush and after the 1906 San Francisco earthquake and fire, the settlers whose business was cutting down trees were often well aware that they were participating in endangering them. A logger baron might set aside a favored tract of old-growth forest for his own family's recreation—or for resort income, as had happened in my neighborhood—while still profiting from the decimation of the ecosystems outside his preserve's boundaries. Many of these private reserves had later passed into the park system, and most of the old-growth redwood groves in California were still named after such men, including Armstrong. It was—and it remained—Kashia Pomo land.

The pioneers' redwood valley was only a mile or two deep; its steep walls led up to the ridges of the Austin Creek State Recreation Area. I hiked up there often. As the terrain transitioned from moist forest to hot hills, the sky opened up and the landscape became voluminous. Habitats alternated between woodland, grassland, and oak chaparral, all in the process of being encroached upon by conifers. Outside the boundaries of the Austin Creek State Recreation Area the land kept on this way for a great many miles, stretching west to the coast, north toward Mendocino County, and east until the Highway 101 corridor intercepted it with the municipalities of Healdsburg and Windsor. In August, wild fennel grew up there. In the dry season, it was hot; wind hammered it. This was the rugged backwoods of western Sonoma County, a keystone connecting small towns, swank Russian River wineries, and the river itself. It was a densely vegetated, difficult-to-access place that touched the edges of a great many populated ones. The Walbridge Fire wasn't the first time those hills had burned, but it was the worst in settler memory.

A week earlier I'd watched videos of firefighters protecting Armstrong Woods. In the old-growth grove the fire had been slow and heavily monitored, and it mostly acted like how fire should act in a redwood forest. It cleaned up the underbrush, burned off diseases and pests, and provided new starts and germination for many native species. But the fire hadn't been slow upslope. In the Austin Creek State Recreation Area, it soared and gusted in multiple directions, at one point threatening to join with another lightning fire on the coast. It flew so quickly that it blew right over some ridges, leaving veins of undamaged oaks in cracks carved into the hills by seasonal waterflow. By the time the Walbridge hurricaned downhill from Austin Creek into Armstrong Woods, this woman's home must have already burned.

She told us she was a widow, apparently recent, and she didn't know if she was up for starting over yet again in the same place. As she recalled her losses, she smiled; there was something wild about her mouth's tilt, almost giddy. She must still have been processing what she'd just seen, but reflected in her eyes was a shadow of distraction, as though she were already thinking of the looming question mark her life had just become.

As Max and the woman talked, my dulled, post-evacuation state insulated me from the intensity of their conversation. I felt guilty that I wasn't particularly curious about this woman's dead husband or her wooden ceiling beams; I wanted to know more about her trees. Which fruits? Were they alive? Would the redwoods resprout? Was the fire still burning underground like they said it could? But I didn't ask.

Did she have a place to stay, I asked instead. Yes, with one of her grown kids in Santa Rosa. But it was crowded; she might try to get a hotel. Did she want a beer, Max suggested. Hell yeah she did. She took the can from him and, momentarily unburdened, wandered back to the water's edge.

On the far shore, two river otters played tag around a kayak upturned on a small wooden dock. Nearby, a solitary pole stood high above the water—the remains of a long-dead Douglas fir tree. Balanced improbably at its top was an impressive tangle of sticks and forest detritus. It was a huge bird's nest, and it had been there ever since Jesse or his mom could remember. The nest belonged to a pair of ospreys, large black-and-white raptors who returned each year from the sea to raise offspring here. In summers the parent birds often circled overhead, hunting fish and guarding their nest from crows and passing cars. Directly below the nest was a sort of platform on which a few branches had fallen; a handful of smaller birds huddled there, likely taking protection from other predators and awaiting dropped fish guts. Behind the nest was the two-lane highway that led to the coast; above that was Jesse's mom's neighborhood; farther up, the ashes that used to be this woman's home.

Standing apart from us, the woman watched the birds and toed the rocky sand and drank her beer. My friends and I became reabsorbed in our own stories and lives. A few minutes later the woman ran back over to us, brandishing the beer can. Have you read this, she demanded.

The marketing copy on the beer can was written in a rambling, half-stoned style that was popular with microbreweries at the time. A run-on sentence extolled the virtues and sensibilities of this IPA, which boiled down to being independent-minded and unafraid of the unknown, or something like that. The blurb ended with a salvo: sometimes you just have to burn down the whole house. It's a sign, said the woman. An omen. She was smiling and the untamed look was back in her eyes. We gave her our phone numbers in case she needed anything, knowing she wouldn't call, and she returned to the birds with her beer.

I lay on my back in the sand and watched what appeared to be dots of dust in my eyes transform into circling hawks beneath the smoky sun. The ospreys guarded their home. The smaller birds guarded the hawks' shadows and gleaned their discarded bones. When the fire crossed the river, and it would, if not this season then sometime soon, this nest would be its last stop before it flew across the shallow green water and continued on.

By the time we got home from the river the air quality chart rainbowed from orange to red to purple. While we had been swimming, the wind had changed. It blew up and down the western seaboard, growing in speed and strength throughout the night. Fires followed wind, and in central and western Oregon dozens of new wildfires were blown into existence that night. As flames gusted through Oregon towns named after other things—Oakland, Phoenix, Salem, Talent, Rainbow—laypeople documented the usual litany of horrors, by now routine yet somehow always shocking. A slow chain of cars caravanned down a burning mountain road. A husband, heading back into the fire to find his family, came upon a woman so badly burned he didn't realize she was his wife. An environmental activist died in the forest he had fought to protect. People ran from flames, drove through flames, were awakened by flames. Fire, wind, fire.

The wind seemed to drive rumors, too: the Oregon governor stated that five hundred thousand people had been evacuated, then corrected it to forty thousand, but by the time that number was arrived upon it had already grown. Far-right groups began circulating intentional misinformation, claiming the fires were set by antifascists and members of the movement for Black lives. White men with guns and ammo belts slung across their chests roamed evacuation zones, threatening journalists on camera. The wind pressed on through the night and into the next day.

Up a thin canyon in a temperate rainforest in central Oregon, one of the areas walloped by wind and fire was the property along the McKenzie River where nature writer Barry Lopez had lived for decades. Over the course of his long career Lopez had witnessed and written about struggling ecosystems all over the world; his book *Arctic Dreams* was a classic of the genre. But Lopez, who was then in an advanced stage of cancer, had difficulty processing the losses of the fire. The trees, the salmon he loved to observe, the roaring river's bank—all were scorched. His writing studio burned down with his work inside. The house wasn't entirely destroyed but it was severely damaged, and he and his wife, the writer Debra Gwartney, had to move to a rental home, where Lopez died several months later. Gwartney later said that in those final months Lopez liked to return to visit the property and walk through the skeletons of the trees. He paced the footprint of his office, she wrote in the literary magazine *Granta,* "a lone sentry at the gates of that phantom building." He dug through the ash and rubble and tried to identify different objects from his life, trying to see them for what they had once been.

· 6 ·

Sky

Different types of matter burned at different temperatures. When a piece of fuel caught fire, it was extremely rare for every part of that fuel to be entirely burned up. There were almost always tiny particles left over. In a fire, those material particles mingled with soot, gases, and water and coalesced as smoke. Smoke was made of what didn't burn.

Smoke was pollution. The carbon dioxide that wildfire smoke created was the same carbon dioxide emitted by the production and burning of gasoline, coal, or any fossil fuel. The dangerous particles of matter carried by smoke were classified as $PM_{2.5}$, a measurement of how infinitesimal and hence how inhalable the particulates were.

I could taste smoke. I could see smoke. I could smell it and I swore that I could touch it. The sound of smoke was less a sound than a dampening, a dome clamped over the sky: snow globe effect. Smoke had a body—it was a mass, a celestial object—but it also could slide down a person or an animal's throat and into the bronchial system and tickle the tiny cells in there. It hurt bodies.

Smoke made the lungs weak, which made Covid easier to catch and worse when it was caught. Smoke was more dangerous for pregnant people and babies. It caused asthma, heart failure, a nasty mid-depth cough that was impossible to source definitively but took years to go away. Smoke was more harmful the longer it stuck around.

Smoke was annoying, tugging and tugging at your peripheral senses until you couldn't concentrate on anything else, except as an object it wasn't solid enough to concentrate on, and so you floated between anger and distraction. It made me feel so tired.

Smoke could create its own weather, clouds, and lightning. It bungled meteorology equipment and algorithms. Smoke could make moths turn color, it could change a person's skin from young to old. It could communicate, its faraway patterns relaying messages; it could travel, sauntering easily from Washington to California, from the boreal forests of Canada to New York City. Smoke always followed beauty, my grandmother said. In the kitchen, smoke could render food a delicacy or a disaster. Outside, smoke could crack open seeds that had been dormant for decades. Smoke was beautiful. It was blue and orange and sallow gray, the color of feathers and annihilation.

In a fire, the black smoke was hotter and the white smoke was wetter.

In a fire, the smoke was usually what killed you first.

In early September of 2020, wildfire smoke was what caused four tule geese, who had been tagged with GPS locators by scientists studying bird migration patterns, to change their flight pattern above Washington and Oregon. They had never encountered the concept of borders before but the smoke created a barrier that now forced them to change their routine. The geese were inborn to travel a particular route and stop at particular places along the way until they arrived, together, at a particular lake in Oregon. The previous year, the geese had flown in what amounted to a direct path to their lake, using the same route and the same stopovers. There were slight variations in how widely they veered off the x-axis of the route—every goose was an individual—but they basically flew in a straight line. In the early weeks of September 2020, however, some flew much higher in order to get above the smoke, using precious energy that required more calories. Some stopped over in unfamiliar places along the way, lost or in need of sustenance. Some aborted the flight and regrouped, waiting for the sky to clear before venturing forward again. But the smoke didn't go away. On a map showing the geese's route, the smoke was visualized as opaque red poufs; the path of the birds was symbolized by a line that wobbled like an ambivalent curlicue.

There was no sense of direction in smoke.

The night of September 8, wildfire smoke rose from British Columbia, Oregon, Washington, Klamath, the Sierra Nevada, Mendocino National Forest, Montana, Colorado, and Wyoming. Smoke commingled and surfed on the air currents. It congealed as one nonsolid mass above the western part of North America, where upon its attempt to descend to the ground in California it was met from below by the marine layer, which was made of particles of moisture coalesced above the ocean and sucked inland by the thirst of valleys each night. Before the smoke could settle onto the land and water below, the fog held it aloft, suspended. Even thus penned in, the smoke generated by the wildfires that week was capable of a presence so oppressive that if a person was standing anywhere on land below that noxious mixture—smoke, fog: smog—they might swear the sky was pressing down on them, heavy, the atmospheric ceiling becoming shorter and shorter until it felt like a restraint. Tethered body, smothered heart. This was how the smoke behaved on the western seaboard the night after I swam in the river. Then in the morning it took away the sun.

The first thing I noticed was that it was cold. Under the trees the hours between seven A.M. and ten A.M. were always a time of great temperature changes, but the morning the sky disappeared, the weather moved in the wrong direction: as the day broke, it got colder.

Then I noticed it was dark, though I'd slept past eight. The light inside the bedroom was amber and grainy. There was a taillights-at-night flicker to its quality. Max was up before me; he came back in the bedroom to snuggle while I tried to wake my body, and he said, So, it looks like the sun didn't rise today.

I had stayed up late scrolling the Oregon fires. My body hurt. What now, I thought. What now, what unrelentingly superlative catastrophe will we be subjected to now?

My pain was tremors, crevasses, abyss.

I had seen partial eclipses but I'd never experienced what eclipse chasers call totality—the complete darkness of a full solar eclipse. The dark of this morning must have been its kin. In my living room, the sliding-glass door framed a scene in which it appeared to be neither day nor night. The redwoods were dreamscapes, umbrous black against a glowing red sky. Although they appeared shadowlike themselves, the trees cast none of their own. There wasn't enough sunlight or contrast to make shadows. From inside, the garden looked two-dimensional; the yard and our neighbors' rooftops and the skyline of treetops beyond had a filmy haze to them that skewed my sense of depth. In the sky, the smoke's akinetic heft made the fog beneath it seem too close. Everything felt thick and totally wrong.

Max went out to the garden to do dawn patrol, and I drank my tea and watched him from the window. Above the smoke, the sun must have been ascending along its normative arc; the sky did grow lighter. But its daytime darkness was replaced by a color I'd never seen before.

On any given day, the human eye can only see certain colors on the spectrum. (A camera lens, even fewer.) On the day without a sky, the marine layer and the unusually large quantities of smoke in the atmosphere above it combined to create a new kind of filter. Some hues were added, some left out, and the light moved through and around all the particles of all the unburned things in the way that only light can. What remained was mostly red. There appeared to be no sun, which made it feel like there was no sky. The usual distinction between atmosphere and earth was rendered invisible. A basic understanding that until today I had taken as unshakable—that there was ether, ozone, the further wilds of the universe above me—was rendered invalid. On this dark morning, anything that was illuminated was tinted by the filter of smoke and made indistinguishable. Just a trick of the light: this extraordinary red.

The only way I had of incorporating into my sensory experience the color of that sky was by comparison, but I had nothing to compare this to. I called it red; others said it was orange. It was not the usual color of sunlight when there was wildfire smoke, which was a sort of rusty gray that I thought of as refracted fire. The sky was not quite the color of a winter fruit—persimmon, pomegranate, the fruits of mythology. It was not the glowing orange that sometimes permeated a certain fifteen seconds of the Pacific Ocean mid-sunset. It had hints of the artificial in its corners, like one of those neon lakes you fly over in an airplane that turns out to be a lithium mining pool. Although it was smog that colored the sky, this wasn't the color I knew as smoggy, which I might define as a fluorescent horizon backdropping a gridlocked freeway during golden hour: luminescent but slightly ominous. There was nothing slight about this red. This was a sky told of fables and omens, cave art and science fiction. The red sky was a sight that might only make sense in a world of wrathful gods, or maybe a world of no gods. It was the color of a sky from classical poetry, a color for high priestesses and jaguar kings. This was a Dante, *Odyssey,* war-begotten red, like dust storms over a burning oil well in Kuwait. It was a color to put people in our place, inside history. And it was impossible to describe.

From the window I took a selfie that included the sky but the photograph failed to capture the quality of the odd, alarming light. It was as though the inner constitution of my phone camera, itself made of tiny pieces of extracted planet, could not conceive of a terrestrial reality in which this color would occur. I didn't share the selfie with anyone. Throughout the day I tried periodically to take more pictures of the sky, but with each attempt the phone adjusted its color settings automatically and turned formidable vermillion into basic orange.

My brain had a few questions. It had accepted rather quickly the illogical concept that the sky had disappeared. But if the sky was gone, then I would need to do some things differently. It was time to water the garden, but should we? Would the plants be too cold or damp in the daytime dark? Had the smoke really blocked out the sun, and if so would vegetation die, would famine follow, might the missing sun somehow skew Earth's tidal rhythms or crack open its core? My brain made these leaps easily despite my intellect raising an eyebrow at them. The sky had disappeared. Anything might occur.

Max came back in, reporting that the garden was fine, it was just really fucking creepy out there. And smoky, he said. But not as smoky as one might expect, considering that the world, by which he meant the western part of North America, was on fire.

This is epic, I said unnecessarily. Max and I both understood immediately that the red sky was a historic event, one that everyone in the region was experiencing whether or not they'd experienced wildfire directly. We couldn't stop talking about it—it's like one of those stories of olden times, I said, when the world goes dark and the people fall to the ground, certain the end is nigh but really it's just an eclipse. I then related to Max the unproven rumor about the Lumière brothers, early filmmakers whose 1896 film of a train heading directly toward the camera had supposedly inspired audiences in France to run screaming from the theater: at once both unreal and too real.

It was a weekday. Work, fire maps, work. I went outside to walk through the roses but the atmosphere quickly felt unbearable, and I came back in. I went online to try to grok the science of what was happening, and to give my brain something to work with, and the pictures had already arrived. The timbre of the sky that day may have been an inexpressible moment, but it was already being conveyed in millions of ways. Every media platform. Every type of exposure. Film, digital, metaphor. It seemed that everyone who lived beneath the sky without a sky was trying to capture it. So many images, over and over, so fast, rendered the red sky both totally monumental and instantly generic. What a cliché—millions of Californians posting selfies while the sky was falling. None of the pictures nailed the color but most of them captured the eerie vibe. The imagery overload blurred my inner boundaries, my sense of order and time. It was as though an errant thumb had been put over the lens of my sensory capabilities. News reports, social media, text messages. The captions all used the same term: apocalyptic. Such hyperbole was forgivable today. Because it felt like the end of the world.

If my evacuation in Santa Cruz had been a slow slide toward catastrophe, the red sky felt like an avalanche. Disasters aside, like everyone alive I knew the general malaise of what had commonly come to be shorthanded as *late capitalism*—the societal and systemic failures that undergirded my conception of the future in ways that I was still uncovering and that imparted a constant, low-key form of despair. And I daily experienced the fears that any anxious brain holds inside itself, the gaming out of every likelihood, everything that might go wrong, as a way to prepare. But the slow, turbid terror that suffused my body when the sky disappeared was something new. It was a deep body fear, an animal truth.

Instead of watching the world end in real time online, I decided to read a book. The previous day I had received in the mail a new version of the eleventh-century epic poem *Beowulf*, translated by fantasy author Maria Dahvana Headley. This *Beowulf* was unlike the poem

I barely remembered reading in my early twenties, not having been assigned it in high school but taking it up later, sometime between my John Donne phase and my fascination with *Buffy the Vampire Slayer.* The plot was fairly simple. A monster, Grendel, stalked a group of noblemen, eating one of them every night. So the humans holed up in the great hall of their castle, hiding from the inhabitants of the land and waters they laid claim to. This went on for a dozen years, nobody did anything, everyone was scared. Pretty early in the book, the titular hero came along and killed Grendel. The people rejoiced. Palace intrigue abounded. Grendel's mother, herself a monster, sought revenge, and eventually Beowulf fought her in her underwater habitat and killed her too. The people rejoiced.

I found Headley's translation of the epic to be contemporary, witty, and fast-paced, and it was refreshingly untroubled by anthropocentrism. In this version, the monster's mother (who was never named but was introduced as a "monster, feminine") became an avenging matriarch, a woman wronged whose violence and incessance were spurred by grief. I found her to be an entirely sympathetic character: a grotesque, made so by the wrongs of men.

My nerves were still on high alert for flight. I curled up on the floor near the sliding-glass door in our living room, which looked out on the garden, and lost myself in monsters for a while. The book was short, and when I finished I looked out at the dark daytime firesky and the shrugging flora. I thought about those humans in that castle in the poem, going to bed afraid every night for decades and every morning waking up to decimation; becoming strangely accustomed to the terror, to the daily experience of loss; pretending that there was a normal way of life they could ever get back to, or that they were any different from the monsters outside.

Under the red not-sky, the ocean pulled me west. Three in the afternoon. I wore the usual two masks—fabric over N95—plus a headlamp. The coast trail traced the edge of a cliff high above the water. Below, colossal boulders bleached white by bird droppings withstood the ceaseless breach of the waves. There were no trees. It was a coastal prairie in which only grasses, coyote brush, and other scrub grew. The tallest things around me were the bush lupines, their flowers long gone, seed pods empty and exploded on the stem. I walked here several times a week; it was one of the only nearby trails that was flat, which was the only kind of hike I could do without guarantee of subsequent pain.

When I began walking on this path years ago, I had at first perceived the landscape to be ordinary compared to the majesty of other Northern California environments. Despite the presence of the Pacific Ocean, the land up on the cliffside was very beige. But the more I visited this part of the Sonoma Coast, the more I came to understand its subtleties. The coastal prairie flitted through its micro-seasons as delicately as the small brown songbirds that hovered on nearby coyote brush when I passed their grounded nesting areas. In different months and weeks, the grasses took on varied shades of green, yellow, brown, even blue. In spring, golden California poppies, purple lupines, and pink cotton balls of coastal buckwheat bloomed on the cliffs. In summer, deer traveled through in large families and rabbit noses poked out from the shrubs at dusk. Year-round, a resident pair of northern harrier hawks cruised over the fields, as did turkey vultures and a pair of ravens whom I was convinced knew me by sight.

In September the grasses were dry and waist-high on either side of the path. I stepped over coyote scat on cracked dirt. The dinosaur profiles of seabirds dotted offshore rock formations. Cormorants air-dried black wings while facing the incoming tide, each in a crucifix pose. At the coast the sky was a less-eerie red than it had been at the house, more of an ombré ochre. Out here, the marine layer was nearer to its cousin the water, and so the smoke too had descended closer to the land, making the air quality worse. The ocean was a languorous dirty brown.

Along with its grasses and vistas, I thought I had memorized the sky in this place. I knew the blues and grays of its moods, the cumuli and nebulae, its surprising seasonal palettes. I had internalized a lifetime of sunsets over this ocean: I thought I could never lose the mental image of how everything went pink, low down above the horizon line, before the sun slipped under and Earth rolled away from it into night, gently collapsing me backward until I could feel the planet's tilt. I thought I had it memorized, until it wasn't there. Now there was just this glowering red.

I walked slowly. There was no indication of the passage of hours from sky or earth. Time felt suspended. After about a mile I veered off on a side path and walked toward a large rock outcropping, one of few geological protrusions on the flat land. There were three or four large boulders, each twenty to thirty feet high, positioned in a sort of circular array with pockets of grassy earth between. Each of the rocks contained an eclectic mixture of geological variegations. This was a popular spot with rock climbers; white traces of their chalk blended with brown, blue, gray, and red stratigraphic stripes. I was looking for a certain nook. On the inland side of the rocks, sheltered from the coast wind, there were several spots about ten feet off the ground where the blueschist rock was unusually smooth and dark. Archaeologists had determined that the rocks were burnished because they had long ago been rubbed repeatedly by the hips and shoulders of mammoths, soothing some prehistoric itch. I located the rubs and reached up to run my hand over them. They were cold. It wasn't hard to picture tremendous beasts rambling through these grasses the way bobcats did now. I lived in that kind of landscape—oversized, timeless. Between the mammoth rocks, at the center of their ancient huddle, there was a small hollow place where the wind never blew. I pushed inside and propped myself up to a seated position on the ground between two boulders, legs on one rock face and back against another. It was still lush in here, sheltered from sun and air. In the crags, some no more than six inches high, miniature landscapes had formed: small ferns grew from stone, frogs mated in unseen puddles below; even the poison oak vines growing there, like talismans to ward off human touch, were miniature. I sat inside this tiny wild place, and I looked up to the bloody sky. I wondered how many times the world had ended out there while inside these diminutive crevasses life continued on.

Worlds ended all the time. I thought of the captions I'd seen beneath photographs of the sky this morning: *End of the world,* they said. *Apocalypse sky. It looks like a movie.* People were understandably struggling for ways to describe the situation, but there was an egotism to such proclamations that I didn't feel comfortable with. The idea of a singular apocalypse suggested there was only one world, only one end. The notion of the end of the world was itself a flexible one: Earth had already experienced five extinction-level events, and it was at the start of a sixth. The past two decades alone had seen several historic tsunamis cause mass death for people living on the coasts or islands of the Indian Ocean, the Sunda Strait, the Java Sea, and parts of the Pacific. Historically, volcanoes had been common world-enders: in the Permian era, about 250 million years ago, a volcanic eruption killed approximately 90 percent of species on Earth. There were all sorts of ways the universe could end you: earthquake, flood, plague. The one with the dinosaurs. And of course the human-made apocalypses: war, genocide, chattel slavery, climate change, the list was as unending as the world. The end of humans had begun the moment someone decided to burn coal for fuel. Or perhaps the root was in one of the many moments when people decided to continue burning it, even though we all knew where this was going.

As countless Indigenous artists and scholars around the world had long been pointing out, the original inhabitants of the Americas had lived through an apocalypse each time new settlers arrived bringing, variously, smallpox, extinction of key species, the belief that land could belong to people, and the belief that people could belong to people. Now Indigenous survivors were living through the sixth extinction along with everyone else. The apocalypse of settler colonialism didn't exclusively kill people; it transformed the ecologies of the places it struck. For nonhuman species, apocalypses happened every day: extinction, migration, extirpation, eradication of habitat. In our relentless quest for oil to burn, Americans and Europeans wiped out most of the whales on the planet in the span of about a hundred years. In North America, the coyote survived multiple attempted apocalypses over several centuries, coming back smarter and stronger each time humans tried to end their world. No wonder it sometimes sounded like they were laughing in the night.

From aboriginal origin myths to Greek myths—which likely began as aboriginal myths themselves, with the Mycenaean people—to blockbuster Hollywood movies and the entire genre of science fiction, ever since humans had experienced apocalypse, people had made culture that predicted our own demise, often via fire. These stories were myths, but the creative minds who produced them often worked from lived experience. In 1883 the cataclysmic eruption of Mount Krakatoa in Indonesia, then called the Dutch East Indies, made the skies in northern Europe appear red and flaming—a harbinger of modernity that Edvard Munch later depicted in the background of his 1894 painting *The Scream*. Earlier that century, the 1815 eruption of Mount Tambora, also in Indonesia, had been the largest since humans began keeping records. The particulate matter that Tambora's eruption sent into the atmosphere set off a series of extreme weather events around the globe that lingered for upward of a year, cooling the temperature of the planet and causing crop shortages and economic chaos in what came to be known as the year without a summer. The apocalyptic sunsets of that year could be seen in J. M. W. Turner's paintings. Tambora had reportedly caused actual climate change, temporarily lowering the temperature of the earth by as much as 1 degree Fahrenheit. Recently a story had made its way around the literary internet that linked Mary

Shelley's *Frankenstein* (fire-related subtitle: *The Modern Prometheus*) to the Tambora disaster: Shelley, then eighteen, newly wed, and grieving the death of her first child, had started writing a monster story while holed up in a vacation rental with her new husband, Percy Bysshe, and their friend Lord Byron during the cold, rainy summer following the eruption. That season's apocalyptic weather—as well as the sights the newlyweds had seen on the tour of famine-stricken Europe they'd taken en route—had helped inspire Shelley's novel of human hubris.

The most compelling version of the Frankenstein anecdote I'd read was in an essay about climate grief by author and academic Sofia Samatar. Samatar juxtaposed Shelley's novel with two poems: the ancient Anglo-Saxon poem "The Wanderer," thought to be from the tenth century, which Samatar interpreted as an early example of writing about solastalgia. Alongside these Western works, Samatar looked at the poetry of Imru' al-Qays, a sixth-century pre-Islamic writer credited with pioneering a common poetry form of the time in which a narrator, traveling through the desert, beholds the abandoned homesite of a loved one. The name of this form, as translated from the Arabic by Samatar, was "standing at the ruins." These three works, Samatar argued, were ultimately stories of grief, each from a different time and culture but all driven by experiences of "extreme weather" and voiced by narrators who "knew world-loss." Humans were in fact a species long obsessed with apocalypse. We predicted, awaited, and worshipped it; and all our prophecies were self-fulfilled.

From inside the tiny wilderness of the boulder array, I heard birds but couldn't see them. I skirted the mammoth rocks and walked back along the cliff path the way I'd come. At a spot overlooking the ocean I observed pelicans flock, using patterns and languages I couldn't understand. I stopped and tried again to take a picture of the sky. I refocused and tapped and zoomed and panned. I switched to video, figuring at least I might capture the boundless sound of the waves.

Normally, the end of the world was one of my favorite stories. When it came specifically to climate apocalypse, the philosopher Donna Haraway, in her book *Staying with the Trouble*, defined two general types of emotional responses: despair—"we're fucked, game over"—or an almost religious adherence to what Haraway called "technofixes," the fantasy that humans can solve climate change with new technologies. If I had to categorize myself prior to 2020, I'd say I was a game-over person. When a real-life disaster happened I followed its news incessantly, to the point where it upset my sleep. When I was a kid, the apocalypse that most obsessed me was Pompeii, the Roman city destroyed and preserved in A.D. 79 by the eruption of Mount Vesuvius. (Pompeii had experienced a devastating earthquake in the years leading up to the eruption; to me, its demise was an earthquake story and the volcano was the sequel.) In my early twenties, one of my favorite books was *Golden Days,* Carolyn See's 1987 nuclear apocalypse novel set among moneyed white women in Los Angeles, which I believed was an under-heralded West Coast complement to Don DeLillo's *White Noise.* Uncharacteristically for an artsy type, I liked blockbuster disaster movies in which comets, geology, disease, or zombies ran amok and brought about the inevitable conclusion to human existence. I felt at home in these nightmarish, if improbable, scenarios. I was envious, even, of the finality a movie apocalypse provided: game over. And as my own world slid more perceptibly into chaos, I took strange comfort in the idea that other people's disasters made my own problems seem small: During my medical crisis, upon returning home from one of my pelvic surgeries I had suddenly craved the light medical gore of *Grey's Anatomy,* a television show I did not admire and hadn't seen in at least a decade. At the start of the pandemic I had been insistent on viewing the nineties epidemic thriller *Outbreak,* which had horrified Max but somehow made me feel reassured. (It turned out the least realistic part of that film was the assumption of a competent and well-intentioned governmental response.) Another favorite rewatch was *Deep Impact,* a 1990s blockbuster in which the closure provided was of the comet-hitting-Earth variety. The hero was, unusually, a woman; more unusually, she didn't have a love story. And she died at the end, which *never* happened. After she was obliterated by a tidal wave, the requisite Hollywood dose of hope was administered by the noble and effective president—played by the noble and effective Morgan Freeman, who seemed to be in all those flicks—but to me the heroine's death was the true climax. It was a silly movie, but it moved me: a woman simply choosing to stay, to be unexcepted from whatever Earth had in store for her.

In 2020 a group of Yanomami people, an Indigenous population who live in the remote central forests of the Amazon, agreed to collaborate with Brazilian filmmaker Luiz Bolognesi to make a new kind of apocalypse film, documenting deforestation of their lands. I later learned about the film from my friends Benjamin and Zoë, who lived half of the time in South America and half of the time in Sonoma County. In the Amazon, there had in recent years been a surge in wildfires. Illegal ranchers and miners—*garimpeiros,* in Portuguese—set the fires in order to clear space for their illicit operations. The illegal ranches and mines poisoned rivers and destroyed ecosystems; the prospectors who worked in them imported violence and diseases to which Indigenous people have no immunity. As Benjamin once told me, what's going on in the Amazon is colonization in real time—not history, not the past, but now. As of 2020, Yanomami territory was about twenty-three million acres, straddling the border of Brazil and Venezuela, although the Yanomami had lived there for many millennia, since long before either of those nations existed. In the 1980s, a gold rush in the Amazon led to more than forty thousand illegal prospectors invading Yanomami territories; in 1986, garimpeiros murdered sixteen Indigenous people, including children, in an act of genocide. After public outcry, the government granted occupation rights to the Yanomami in the early 1990s, and the rapid pace of extraction temporarily decelerated. But when fascist president Jair Bolsonaro came to power in Brazil in 2019, he issued what amounted to an open invitation to industry: Come back to the Amazon, he indicated. Do illegal mining and ranching. Cut and burn the trees, take the land, harm its people, I won't stop you. And they came. It was within this context that the Yanomami partnered with a documentarian to make *The Last Forest.*

The Last Forest was unlike any apocalypse movie I'd ever seen. It was technically a documentary following daily life in a Yanomami village. From the beginning, the director, who was an outsider, understood that using standard vérité and talking head methods of documentary filmmaking wouldn't work with the Yanomami; these techniques were too reminiscent of anthropological incursions that had traumatized generations of Indigenous people. Besides, to convey the truths of Yanomami culture to outsiders, the story needed to be told in a manner that was true to Yanomami culture. The community agreed to collaborate with Bolognesi in telling their story, their way, and Yanomami shaman and land rights activist Davi Kopenawa shared screenwriting credit. The resulting film switched deftly between documentary, drama, and fantasia. The subjects were also the actors, and many scenes crossed what I, a North American of European descent, might call the line between history and mythology. The result was a documentary in which Indigenous spirits and gods were not symbols; they were characters, portrayed by actors from the community.

One plotline in the film involved intercuts of both a dramatization and an oral telling of the Yanomami creation story, in which two brother gods created the rivers in the forest, then one of the brothers created humans by mating with a fish spirit woman. Eventually, the other god-brother turned bad. He raped the fish spirit, and was banished to the other side of the river, where he buried evil spirits and ores in the earth. Thereafter, anyone who exploited the minerals below the soil would unleash that evil and release the smoke disease—an apocalypse explicitly defined in the film as inclusive of respiratory illnesses such as tuberculosis, pneumonia, and Covid, all of which still disproportionately killed Indigenous people throughout Brazil. To the Yanomami the forest was home, but it was also where all water, all life, flowed from. Protecting the forest was not only a sacred duty; it was the reason for human existence.

At the coast, above the cliff, the sky swirled, smoke and cloud locked in a tug-of-war over the atmosphere. The water eddied. I looked out at the new texture of the planet's ceiling and all the worst-case scenarios of recent weeks churned inside my mind's eye then evaporated. They were already expired mythologies, in many ways, images of the ruination of a civilization that had never been all that civilized. The day without a sky was pushing me further into a wildness, or maybe it was more like a worldness—a place of fire and sky where no space existed between them, and god-men and monsters no longer held sway.

Donna Haraway's book was heavily influenced by the fantasy author Ursula K. Le Guin. In the 1980s, Le Guin had developed what she called a carrier bag theory of fiction, which riffed on a then new anthropological theory that the earliest human tools were vessels—gourds, jugs, or bowls—and not spears as previously assumed. Le Guin found new possibilities in the likelihood that early, so-called primitive human cultures were not militaristic but rather communal and caring. She applied this discovery to the idea of storytelling. Traditionally structured Western stories generally went like this: a hero went on a journey and then triumphed over evil, saving the world (or whatever person or place symbolized the world to him). Le Guin called this type of narrative model the product of a *spear-centric* culture that was by definition also a phallic one. All that thrusting and parrying, all those explosive climaxes.

Any story based on salvation necessitated the prospect of its opposite, apocalypse. Le Guin believed the relationship between people, our culture, and our planet didn't have to be one of catastrophe or salvation. Instead, the stories people tell ourselves about that relationship might become containers for abundance, vessels in which we could carry ourselves and each other through whatever came next. Such a shift in perspective required a reimagining of both the past and the future, an understanding that, as Le Guin wrote in her genre-defying postapocalyptic novel *Always Coming Home,* "all we ever have is here, now."

Always Coming Home was set in Napa Valley, where during her childhood Le Guin's family had a summer home. (Her father was Berkeley anthropologist Alfred Kroeber, whose sometimes controversial work on Indigenous cultures influenced her worldview.) Published in 1985, the novel was structured as a collection of artifacts of the Kesh people, a future civilization that "might be going to have lived a long, long time from now." The book was nothing if not a vessel. It contained, in no particular order, Kesh creation stories, maps, plays, a dictionary, music, and fragments of novels. The structure allowed—or forced—the reader to become an archaeologist, interpreting the ruins of a society that would begin long after their own failed. The premise hinged on readers' implicit understanding that the end of their world had already happened, but the world had hardly ended there.

As I watched the ocean bubble beneath the particle-filled sky, I wondered how much of my fear of these fires, that smoke, was really about the world ending and how much was about *my* world ending? I wanted to be able to distinguish between them but I wasn't confident I could yet tell the difference. Before the day without a sky, I had known I was living in an era of climate change. Since childhood I had been told that climate collapse would happen within my lifetime; I was never in denial about that. I understood that because of economic inequality and environmental racism, the crisis would hit other people first: residents of the global South, Indigenous people, people who relied on land directly to live, and those who were underserved by infrastructure in rural and urban areas. But I had ended my scenario there. Because I hadn't yet fully accepted that I, too, lived in a *place* of climate change—and because on many levels I was still clinging to the story of salvation through exceptionalism that I'd struggled with during my beachside evacuation in Santa Cruz—I had left myself out of my own apocalypse. This was the situation I found myself in, after four decades of experience living as an American in what academics variously called Anthropocene, Capitalocene, Pyrocene, Plantationocene, Chthulucene times: I had a surplus of world-ending visions, images of blazing skies and big finishes. But for all my Cassandrian confidence in the decline of *Homo sapiens,* I had little capacity with which to navigate the everyday experience of living inside it. It was like bringing survival supplies to a beach house, prepared to splint a broken leg or claw my way through rubble, but forgetting to pack toothpaste.

When I got home from the coast I put a video clip of the amber ocean on social media, my poorly adjusted visions adding to the ouroboros of everyone else's, each an attempt to capture something, anything, everything. Soon the weather grew warmer, which on this upside-down day I took to mean night was approaching. Max and I must have eaten, talked, done the usual home things, but the evening faded from memory as soon as it was lived. Most likely we spent the hours between the dark day and the dark night on computer screens, participating in the great, perpetual scroll. It was as it had been: every piece of bird's-eye footage of every burned neighborhood, every shot of the San Francisco skyline backgrounded in red, every bell curve graph or cruel word out of a cruel politician's mouth—they each felt like impossible stories to tell.

I eventually clicked my way to a stunning picture, the one image of the day that made me stop and look closely. It was a satellite picture of Earth taken earlier that morning. In the photograph, the blue planet was a perfect circle. But something askew immediately caught the viewer's eye: there was one color that stood out against that Pangaean palette, discordant. At the border between an ocean and a continent, beneath a white fuzz of jet stream clouds, a brown funnel seemed to erupt from a pinpoint too small to see. There was another near it; another; they were multitudinous, and the off-hue streams they emitted billowed and coagulated and blocked from view the land below.

I looked at the picture of Earth for a very long time. In future months I would now and then find it saved in my computer and gaze at it anew. It was terrible and beautiful; it was a picture of what was happening, not a fiction or a fantasy. It was the western coast of a place once called by some of its inhabitants Turtle Island, now known as North America. It was a picture of California, fire, and me.

· 7 ·

Smoke

The blur of the sky extended into a fuzz of days that felt indistinguishable. By early September, the world had passed the benchmark of one million deaths from the coronavirus. The fires up and down the Pacific Crest burned. The smoke smothered. Air quality indexes soared and calmed along with the wind. Newspapers published stories declaring the air in major West Coast cities to be the worst in the world, an unusual distinction for a United States locale despite the country being responsible for the most per capita carbon dioxide emissions globally. The president chided those who correctly traced the new, bigger wildfires to climate change and promised the nation the climate would get cooler soon, *you just watch*. The weather continued to be very hot.

In the small white house under large red trees, there were days when the breeze pushed the smoke entirely away and allowed a breather. Max and I fell into a reactive pattern as the month wore on: When it was smoky, we did inside things; when it was clear, we went outside, urgently, desperately, like children at the start of summer break. Beach walks, forest walks, long hours spent in the hammock, drinking sunlight. Mostly, outside days meant the garden.

With the help of a local landscaper whom we paid $30 an hour, Max continued clearing the wilder parts of the hillside in the yard, hacking away vines and bushes on the steep slope below the flower garden. At the far bottom of the yard, in a little valley between the fence and the slope, we put lawn furniture. That's where the fire pit will go, Max said. Haha, no, I replied.

Cleared of decades of neglect, the hillside took on new aspects. The soil, now visible, was red brown, indicative of clay. Modesty, a native ground cover with delicate white flowers, was brought to light. I knew there must be so many more plants being choked out beneath the overgrowth—wildflowers, clumps of graceful sweetgrass, and rangy stands of huckleberry. When the fire comes here, I thought, the soil will be laid open to the sun again and they'll grow back stronger, erasing my impact.

A WEEK AFTER the red sky, mid-September: The light was no longer world-destroying orange. Now it was a more familiar fire-season gray backed with warm tones. I stood in the garden clicking my pruning shears open and closed. I was in pain and ignoring it. It was afternoon. The sun was obscured enough by the smoke that I could look at it directly, which I did, then regretted. The trees were still a bit blurry. I noticed that, for the first time since the day of no sky, the redwoods were casting shadows again.

In my calendar there was a scribble: *do summer pruning.* The note moved along with the rest of my to-dos from weekly spread to weekly spread as emergency conditions persisted. Whereas winter pruning of deciduous fruit trees encouraged growth, summer pruning limited it. The right time to prune was after terminal buds had set on the branches, but before the leaves dropped for winter. September was late for this task, but luckily for me summer never ended anymore, and the fruit trees were far from dormancy.

Winter pruning required faith; summer pruning took imagination. I approached the Wickson crab apple, a strong tree with fragrant blossoms that made surprisingly sweet fruit. I stood back and tried to visualize the shape I wanted the tree to take in the future. Like a sculptor creating form out of negative space, during summer pruning the gardener must ask herself, What parts of the tree *don't* I want to grow? It had an open-center form; sun cup, I reminded myself—my dad's shorthand. With this type of tree form, as opposed to the classic Christmas-tree shape, the tree looked more like a vessel, open in the middle and ready to fill with sunlight.

I had in hand a notebook with sketches that I'd made of each apple tree before the evacuation sidelined my horticultural vigilance. I had drawn bold little lines where I intended to cut. I needed these visual aids; I still didn't fully believe that what I did in the garden would have a positive effect, a notion born partly of lack of confidence and partly of the earned knowledge that despite my labor and worry, I had little say over what happened to these plants. When I lived in cities, I used to think that gardening was about convincing things to grow, coaxing new life from dirt. It seemed difficult. At the house in the woods, however, the forest around me and the garden I set about growing within it gave proof that most plants didn't need much coaxing. I found that my flowers and fruit trees grew easily, rapidly, too much, really. As gardener my role became more about limiting the way in which plants grew, manipulating their form, and sometimes beating them back if needed. All those interventions.

Today I located the scar tissue from the prior year's pruning in the Golden Russet apple tree. I shouldn't prune lower than this line of division on the branch. The two-year-old tree was a central-leader form: Christmas tree. At its top, two shoots vied for apical dominance. This tree had an unbalanced structure, probably from its early life clumped together with other bare-root trees at the nursery; most of its branches were on one side. To remedy this, during winter pruning Max had taken a knife and made little half-crescent cuts at strategic points on the bare side of the trunk, a technique known as notching that can sometimes make new branches sprout. It had been successful; in two of the notches, new branches grew from the scars, each only about ten inches long. Those I would leave untouched, as I didn't want to limit their growth yet. I stepped back to start visualizing my negative space. I imagined the tree covered in abundant new growth. For a moment I envied the tree's resilience. Like a surgeon I cut swiftly and with pleasure. In the black oak an acorn woodpecker laughed.

The interventions on my body were ongoing. Today the pain was lightning again, tiny violences at random intervals that caused me to stop and catch my breath each time they struck. I had done multiple rounds of pelvic floor physical therapy, which consisted of the therapist manually accessing muscles in my deepest places. I had tried biofeedback, gentle electric shocks, which offered some initial relief but was not covered by my insurance, so I had to stop. My new gynecologist, the good one, the only good one I ever found, suggested I might be a candidate for an experimental treatment in which Botox was injected into the pelvic floor muscles in order to break the cycle of muscle spasm, but Max's work insurance changed, and I found myself no longer within her network. I once paid full price to visit a Berkeley MD who practiced what she called deep-needle acupuncture, her own questionably interpreted derivative of traditional Chinese medicine that turned out to be the placement of foot-long needles into my pelvic floor that caused my pain to flare ferociously for a week and left my abdomen covered in avocado-sized bruises. I regularly rotated through trials of muscle relaxants, painkillers, anxiety medication, nerve pain medication, cannabis, psilocybin mushrooms, herbs, and alcohol. I did acupuncture, craniosacral therapy, yoga, Ayurvedic herbs, something called Mayan belly massage that actually helped, and elimination diets. My insurance changed again; I got the Botox; the effects were unclear. And I sometimes went months without doing anything at all, sick of appointments, weary of being poked. Since the start of the pandemic I had by necessity been poke-free. My body hadn't seemed to notice. It would ultimately do whatever it wanted anyway, just like a tree.

As I grew into the realization that the pain might never go away, and my body would certainly never be the same as it was before my surgeries, I also grew more at ease with experimenting in the garden. I still marveled each time a flower bloomed, each time the fruit trees didn't die from the repeated assault of my hands. But as my body's interventions carved their circular path through my life, and as the garden grew and the forest's bramble came bounding back toward the flower beds every season, I had come to understand this viewpoint as paternalistic. The plants weren't weak or innocent; they were far stronger than I was. We had evolved together over millions of years, plants and people; the need between us was mutual, and we each had a role to fulfill. My dad's favorite teaching bit was to compare the relationship between plant and person to that of a jazz call and response. First, you do something to the plant, he'd say, then the plant responds. Then, you respond. *Da capo:* repeat. When I decided to form a garden in collaboration with the land around me, I had joined the band. The agreement was I would give the plants nutrients and water. I would cut them and manipulate their bodies. The things that I did to them would increase their odds of survival. And they would essentially do the same for me, giving me beauty, sustenance, and joy. The palliative reactions of my mind, heart, and body were side effects.

As I settled into a new identity as a person in pain, chronically, I found that to my surprise I had become what my dad called a whacker when it came to pruning. (The opposite of a whacker is a haircutter: just a trim, a little off the top, nothing drastic. Ideally, a gardener lands in between those extremes.) Max put it more bluntly, although with a smile: I was a killer in the garden. I could sometimes be ruthless. If something didn't please me, I moved it. I deadheaded daisies before they were ready and robbed roses of new buds for my bouquets. I'd uproot perennials when they failed to expand at pace. After my early timidity when it came to pruning, it turned out that I wasn't into gardening for the nurturing. I liked the parts where you fucked with the plants. I liked the fact that they responded. And I liked how little they ultimately seemed to care. In the riotous relationship between humans and other organic, living bodies, in the exchange of care, I had become mesmerized by the undeniable, clichéd brilliance of a thing continuing to thrive despite repeated assaults on its structure. Max was kidding, mostly: I wasn't really a killer; I had been known to cry for a dead seedling. It was more that I was becoming unafraid to act. Since those early nervous seasons, I had become a better and more assertive partner to the flowers and the trees.

I paused pruning to do a little training on the Ashmead's Kernel, another good cider apple. I tied branches to stakes with strings, stretching them into more ideal alignment. They should be like spokes on a wheel. From a few vigorous branches I hung weights I'd made at home by filling Dixie cups with concrete mix. The weights would pull the main branches downward, discouraging further upward growth and instead encouraging lateral fruit-bearing branches to form. I was asking these limbs to stretch in directions they would never normally seek. Over many months and years, I would try my best to listen to their response. The other components of this great dialogue—sunlight, water, pollen, dirt—were largely up to them to discover.

I moved on to the next tree and resumed whacking. You have to be careful with summer pruning, my dad once told me. As a newly confessed killer, I was still a little frightened at how far I might take it, how much hurt I could inflict if I so chose. I kept cutting anyway. I found a rhythm, and the slice of my secateurs around the tender arm of each shoot was oddly satisfying. I would never stop being amazed that the relationship humans had developed with the natural world could be so violent and yet so full of care, but I wasn't quite sure where in that relationship I played a role. There were times when I felt like my pain brought me closer to plants and animals than to people. The fight-or-flight instinct never turned off. My body sensed danger and my animal body reacted. *Da capo*.

One of my favorite garden writers, Masanobu Fukuoka, advocated for a more laid-back approach to the relationship between plants and people. His 1975 book *The One-Straw Revolution* outlined his philosophy of hands-off plant tending, which he termed "natural farming." The basic principle was to try to set the ideal conditions for crops to grow, and then to intervene as little as possible.

Fukuoka's parents had been citrus farmers, but Fukuoka instead became a scientist. When he was a young man, in the 1930s, his health had played a role in drawing him back to the outdoors. Fukuoka experienced a terrible bout of pneumonia, after which he suffered a breakdown of sorts and took to wandering the countryside. One night on a hilltop he collapsed, exhausted, and had what he described as an epiphany. The phrase with which he summed up this revelation was "There is really nothing at all." While out wandering the hills and forests, he had realized that nothing he had been placing value in really mattered. What did matter, very much, was that "the leaves danced green and sparkling," he wrote. Humans, by comparison, knew nothing. We were nothing. It was a joyous negation.

Fukuoka then decided to move back to his parents' citrus farm and take up the family trade. At first, he wasn't going to fuck with the trees at all. He wasn't quite as into jamming out with plants as my dad was; he believed that *any* cut was an injury to a tree. And so he didn't prune. But humans had already intervened in the natural pattern and growth of these trees. Left without the care to which they'd become accustomed, his farm's trees did poorly. Their leaves grew too close together, beckoning fungus and insects; fruit became sour, reverting to principles of reproduction rather than flavor; the trees' forms grew haphazardly and became poorly suited to capturing sunlight. Many died. There was a difference, Fukuoka was learning, between "non-intervention and abandonment." By refusing to lay a hand on the trees, he had set off an endless cycle of "disorder." And so Fukuoka conceded to tending his trees, but he tried to do so minimally, in ways that pushed them gently toward self-sufficiency. Fukuoka believed it would have been better had humans not intervened in the first place. But since we had, we now had a responsibility to uphold our side of the arrangement.

Prune to listen. Prune to care. Prune to promise. I tore into the Santa Rosa plum last. It was a beast, a supposed dwarf rootstock but only a year old and already twelve feet high, too high for me to safely reach the top of the leader branch to trim it. I decided to replace the leader branch entirely, cutting it at the base and allowing a shorter, sturdier branch to take the lead. This should shift the tree's focal energy to a less-vigorous pacesetter. When I cut the leader off, I listened carefully to the way my clippers seemed at first to resist closing around the branch, before snapping closed with a slicing of their blades. It was an addictive sound. It was power. As the power went to my head, I had to remind myself that a garden wasn't about controlling nature any more than it was about purity or perfection. It was always messy out here. Disorder ruled the day. I wasn't standing in 250 AQI urging plants to grow because I thought I had the power to control what happened to either of us. I gardened as a call, seeking response. I gardened in curiosity and responsibility. I did it because it's in my nature.

The smoke came back, and the daily oscillation between confinement and escape continued. On inside days, I struggled against the feeling that my life was being enclosed. Work, smoke, work, screens. The wildfires remained distant but extant; containment levels slowly crept upward. Since moving to Sonoma County, Max and I had somehow become people who canned things: One sweaty night we boiled and sealed twenty pounds of tomatoes from our CSA, bottling the taste of this strange summer. Another night, stir-crazy, I pulled out my sewing machine and finished my perpetual pile of mending. I was getting into audiobooks. Max was learning mandolin.

The day Toots Hibbert died of Covid it was an inside day. My upbringing had included listening to a lot of reggae, so much so that I'd sworn off it for several decades. But recently Max and I had been getting back into the classics of the genre, as part of a 2020-inspired effort to be less stressed. Reggae vibes, we called it: I think it's time for reggae vibes, I'd say, and we'd slip a Wailers record on and dance around and I'd be sent back to the glow of my hometown's sunshine. The voice of Hibbert, leader of the ska/reggae band Toots and the Maytals, had been one of my earliest dance partners. As tribute, we put *Funky Kingston* on the turntable and danced. I shook it all off. Max took my hand and pulled me into a spin, and out again, and Toots sang himself into the pantheon of this pandemic's dead.

The record stopped. Max flopped on the couch. The reggae vibes having dispersed, boredom returned. I found myself craving deeper cuts, rhythms and cultures that felt to me so ingrained in my body, even my pain couldn't reach them. Living room dancing always reminded me of my childhood choreography sessions; I missed Rhys. I knelt down and began to flip through my chaotic collection of records, which were lodged in no particular order on a long, low redwood shelf. Summoning old skills from years spent working retail, I instinctively began to alphabetize the vinyl, making little piles for different sections of the alphabet: *A–E, F–K,* and so on. *S* got its own pile because band names always started with an *S* for some reason. I was just trying to keep busy in the interminable smoke, but as I shuffled through the vinyl I was transported into a cloud of nostalgia for my own life. In all this music were the traces of my past identities and eras. *Z* was for John Zorn, who often played at the New York City jazz club where I worked after dropping out of college; I used to smoke cigarettes inside the box office and sneak Elliott Smith CDs into the bar stereo. The warped heft of a Buzzcocks album embodied my Portland era, when I worked at a record store and punk cred was a prerequisite for the job; the tripped-out Brit-gospel of Spiritualized was the soundtrack to late nights after closing shifts at the store, listening to records with a coworker I was secretly (I thought) in love with. Sam Cooke had wandered with me alone in Paris, heartbroken, at age nineteen. Every Bob Dylan record sent me directly to the wayback of my dad's car in my childhood, nobody talking and everybody listening. Ella Fitzgerald and Louis Armstrong were the soundtrack to teenage weekends at Rhys's house, dancing and sliding around in socks on the kitchen floor with her dad, who died young. In the *F* section there was a baby riot grrl, cutting high school to go to a Fugazi show in the city and blaring hip-hop in Sara's mom's truck on the way there. Just the tilde on the name *Gilberto, João* could summon my mom's reedy soprano coming from the armchair where she graded papers at night. The distinctive blue of Coltrane cover art, the pink photo collage on a Muscle Shoals soul compilation, the delicate slide of a vinyl disc into a plastic sleeve: these were all sensations that could be traced to after-hours sessions at the bookstore where my friend Brooke and I worked in high school. With the lights low and the doors locked to customers we sat on the counter and soaked up music recommendations from our older male coworkers. One of them told me my name sounded like a Velvet Underground song: he'd sing *Man-ju-luuuh* to the tune of "Heroin." Wilco's woeful alt-country encapsulated the experience of lying on the floor in San Francisco in the early 2000s, depressed. Later, in the days of young love, on that same floor I kissed Max again and again over the duration of the Magnetic Fields triple album *69 Love Songs.* There were so many people and places contained in these objects. If they burned in a fire, the circles of plastic would melt into one, creating a useless composite whole of me, but I'd still know all the bridges and verses by heart.

It was approaching the autumnal equinox, and also my birthday. After Max and I went through the now-customary Covid/fires decision-making process—any gathering had to be outdoors, but what would we tell our guests if the AQI was above 100, or 150, or what if the wind changed?—the elements came through for me and the smoke backed off enough to make a gathering possible, if not ideal.

Directly below the coastal trail I liked to hike, several sets of wobbly wooden stairs led down to a small cove. The beach sand there was almost black. It had been transformed so recently from rocks that it was more a conglomeration of pebbles than grains. Here the air was cooler than at our house. The sun was out, but its light was tainted a weak, Kool-Aid orange. Max's sibling, Sal, and their partner drove up from Oakland. So did several of my oldest friends. (Rhys couldn't make it, she was working.) Jesse and his mom came, and a few newer friends from the North Bay. We were artists, organizers, teachers, and students. Most of us hadn't seen each other in at least six months. The party attendees wore masks; on the sand, we placed our blankets six feet apart. Max mixed drinks in paper cups and passed around slices of the huge cake he'd ordered for me, chocolate-raspberry from a local dive bar that had started doing takeout during the pandemic. The mood was subdued, the stresses of the moment infusing everything with a sort of quietude, but it was a party.

Toward sunset I sat on a quilt alone and picked at sand flies as they jumped on the cotton squares. Around me, one of my oldest friends talked about prison abolition with one of my newer friends. I watched my friend Kara's toddler run up one side of a boulder, jump off it into the sand, then declare that she wanted to do it *again, again*. Her mom nodded wearily. As the game cycled, the child's swan dives grew higher and bolder, her mother less accommodating, but the kid remained unfazed. She was fearless, Kara said, half apologizing and half bragging. I said I'd start thinking of roller derby names for her.

I gripped the blanket between my sand-dirtied feet. I had painted my toenails red on an interminable inside day. In the weeks since the day without a sky, everything I looked at seemed to take on a red tint. Someone had made an altar to autumn by wedging into a granite rock a shaggy bouquet of red-leaved plum and maple branches. In the sand next to me, the bright chemical red of my abandoned Aperol spritz glowed neon in its white cup, and that small medallion of color contrasted pleasantly with the black-shell sand. The water was low and still today; the horizon wavered like a mirage. I could smell the salt in the ocean for the first time in weeks.

I looked around at my people. I had wanted a party because I wanted to lighten the weight of my worries, the anxiety of the era. I didn't feel lighter, but the weightiness I felt was not unappealing. We were solid, the heft of these relationships holding us together. We could carry a lot.

The toddler jumped off the rock again, again, and a small cluster of pelicans sauntered by overhead.

At dusk Max gathered everyone in a semicircle at the water's edge and I passed around breadcrumbs from a stale baguette. Max explained that this party, in addition to celebrating the forty-third September equinox since my illustrious birth, coincided with the start of the High Holidays. Tonight was Rosh Hashanah, the Jewish New Year. In the Jewish tradition, it was said that at this time of year the books of life and death were opened, ready to take names; as I understood it, everyone would end up in one book or the other, but we didn't know which one. The books closed ten days later on Yom Kippur, a day of atonement and fasting on which adherents wore white and abstained from life-giving activities, including food, water, and sex. In synagogues, the prayer services lasted all day. The time between Rosh

Hashanah and Yom Kippur was one of reckoning, in which people took responsibility for the harm they had caused and the ways in which they had failed in the past year, preparing for whatever came next. These were called the Days of Awe.

Max invited our friends to perform a take on the ritual tashlich, or casting off. The object of the game, Max said, was to toss your breadcrumbs into the ocean and ask the waves to carry away with them whatever was burdening you. Many traditions and religions had this type of a ceremony: a taking-away by water, a symbolic restart. Songdance of birds, renewal of tides.

Think about what you want to unload, get rid of, this new year, Max said. It can be anything; but you have to say it out loud.

I hadn't known beforehand what words to offer in the ritual. I usually hated these types of things—group assignments where you had to enact some public form of confession, go around the table and say what you were grateful for. It wasn't my style. But as I looked around the beach I realized that I had been afraid to hug these people for six months now. Fear was everywhere—fear of the virus, fascism, economic precarity, climate chaos. Fear of fire. I was afraid of sleeping without car keys by the bed or leaving the house to walk in the kindling pile that was the forest up the hill. I probably couldn't unburden a lot of those fears; they were circumstantial and not unwarranted. But there was that other fear, the under-the-surface culpability that had been cropping up more frequently since the lightning fires began. It was sometimes called the middle-class fear of falling. And it was definitely a white American fear, of losing my status, my power. The fear manifested in me as an amorphous foreboding, a discomfort with being subject to the planet and its miracles the same as everyone else was. I knew it as slippage. I'd learned that the way my body dealt with this feeling was, reflexively, to try to protect me—duck and cover. Clench and stab. But as any toddler knew, the best way to survive a fall was to stay loose, to hit the ground rolling. How would my fear change if I stopped apocalypsing and tried to embody this moment in collapse?

In Donna Haraway's paradigm of climate change denial, game-over people rushed to a conclusion of despair, while technofixers doubled down on the false hope that humans would fix this, despite daily evidence that those with the power to do so weren't even trying. Haraway proposed a third option: "staying in the trouble." It was a simple but powerful concept: only by allowing oneself to exist within—to fully feel—the terror, sadness, and strange beauty of what was happening could one develop new ways to live through it. When the early poet Imru' al-Qays, chronicler of climate grief back when it was simply called grief, came across the ruins of his own world, he stood with them for a while. That's how he began to endure the loss. Only by first allowing himself to experience the damage could he then carry on and create from the wreckage a new form, something like a song. As with the Yanomami's hybrid filmmaking style, the messiest parts of the story—the places and moments where the line between life and death, myth and documentary blurred—were the most transformative.

I stood with my friends at the tide line. I shifted my weight from foot to foot. The tug of my pain eddied below my surface, reminding me that I had plenty of experience sitting inside ruination. To move through pain was to be, perennially, in trouble.

On the beach—the temple of temples—the ocean crashed and resonated as it always did, under skies heavy with the smoke of the world we'd made. Max counted off *one two three,* and everyone threw the bread and yelled their burdens at the same time. I aimed my sourdough crumbs in the direction of an offshore rock, on which three harbor seals napped, and I whispered: that fear. The cacophonies of our vocalized troubles mixed with the white noise of the tide and rendered each of our words indistinguishable. It sounded like celebration, I thought. And why wouldn't it be? We were all linked to the problems of humanity, but these people and I were stronger for our linkage, sweeter in our shouting. A new year was coming. In the bluing dusk the white pops of bread floated in foam on the gray waves. They looked to me like sustenance, not like burdens.

We weren't religious. Max had in fact been raised actively anti-religion. He was half-Jewish on his father's side. His family, like many Jewish families I knew, was more interested in cultural and intellectual Jewishness than worship; they were avowed atheists. To Max being Jewish meant being a socialist; Jewish praxis meant questioning everything, and also arguing about it loudly as a form of recreation. But after a psilocybin-fueled revelation in the Sierra Nevada the prior summer in which his ancestors were heavily featured, Max had bought a copy of the Torah that was annotated with progressive commentary by women rabbis. In the early months of the pandemic, when life felt so biblical anyway, he had begun to read it. On inside nights, he curled up in his grandfather's Eames chair next to the brick fireplace, smoked a little weed, read, then told me all about it.

I didn't partake of religion either, but it turned out that I loved talking about the Torah with Max. This, I thought, was the ideal way to receive spiritual teachings: someone else read the big old boring book and then, a little bit buzzed and a lot reinterpreted, related it back to me with a socialist-environmentalist, liberation slant. We'd discuss each passage and wrestle with the hard parts—the slavery and sexism, and the boring bits about construction or cattle. As we debated those ancient happenings, we also discussed current ones. Max had family in Israel, and when I went with him to visit I had been shocked, though not surprised, by the apartheid and racism I witnessed as part of the state of Israel's occupation of Palestine and violence against Palestinian people.

My parents had raised me to be politically aware, recognize injustice, and mistrust dominant systems. They had both been raised Catholic, although my dad once told me his mom had sort of been excommunicated for marrying my Protestant grandfather, I never got the full story. For both my parents in different ways, childhood religion had faded and rebellion had taken spiritual hold of their lives. Then came yoga. I had only cracked open a Bible twice: once as a child in a hotel room where upon declaiming *this book is sexist* I shoved it back into the bedside drawer, and once in high school as an assigned companion text to *Moby-Dick*. After my parents left the spiritual community in which my brother and I were born and named, my dad never returned to any organized form of belief; he found his gods in the garden and the soil. My mom, however, never seemed to stop looking for something to believe in. Throughout my childhood she followed various New Age spiritual directions. She did dream work, Jungian archetype work, color work; there was always a new revelation to be had about how the universe was organized, a new practice to be implemented. I tended to roll my eyes at her constant searching, believing her mission to be an unattainable one. Like many people of my generation, my religion was being against things, counter the dominant culture, underground. But even in my Gen X cynicism I had to admit I had inherited from my mom a strong conviction that people were not the most powerful or important things in the universe.

A new year was a chance to start again, cut old growth to the ground, and begin to formulate new extensions of the self. I took it seriously. The book of life opened, you atoned, the book closed; you were in or out. Then the cycle started anew. The world only ended in order to begin again. You didn't get one without the other. The rituals performed on Kol Nidre, the evening before Yom Kippur, involved reciting a lengthy incantation—a litany of one's misdeeds for the past year. I had gone to the ceremony once at the progressive synagogue in Oakland with Max and some other Jewish friends. Although

I demurred whenever the lyrics called for me to say the word god, I had found it a surprisingly moving experience to chant, out loud, the things I had done that were harmful, from the misdeeds of my own relationships and ego trips to the larger commissions of the society and species I was a part of. As they sang their wrongs, people pounded their chests. It was intense. But this year the services were on video, and I just couldn't, so Max and I opted instead to go outside. We'd pay our dues to the landscape directly and skip the litany. The year 5780 had been difficult enough; why beat myself up about it?

It had been a week since my beach birthday party. The AQI was yellow that evening. The day had been sickly hot. My pain was magma, a slow subterranean boil. The trail we picked was easy, a gentle up-and-down slope for about a half mile beneath the shade of redwoods, firs, and bays. It broke out of the trees and dead-ended in a spectacular view: a large creek's watershed ran from this park to the ocean, ten miles west. Between me and the Pacific was the valley's diminishing horizon point, a descending vee of green conifer ridges that seemed from my perch to be almost within arm's reach. Geography felt abbreviated here.

A watershed when seen from above was so obvious in its shape and purpose. In the cut of the riverbanks and the curve of the valley I thought I could see the vestiges of those mighty prehistoric rivers, the wellspring of the High Sierra, culminating at my feet. Directly below me, young madrone trees dotted a parched hillside between bushes of ceanothus, California lilac. Between the bushes the soil was bare and red. A month of persistent smoke had desiccated everything, and the ceaseless wind chapped the last soft parts of the land. I recalled ragged Sierra peaks, lush glacier-fed valleys. Where was that water now, when we needed it? I wanted rain to put out the fires. I wanted fog to halt the wind. I wanted it green again, I wanted it green forever. I wanted to lift the ocean before me and pull it inland toward the disappearing chaparral and the dying oak trees and the thousands and thousands of people, afraid.

I figure we have probably anywhere between two days and ten years before our house burns down, I said wryly. Max didn't reply. I beheld the sweep of the watershed and imagined it a river of fire. This was Pomo land.

Did Richard tell you they closed on the house? I asked.

Since the lightning fires had started, people we knew had been decamping. Richard and Jay, who lived next door to Frank, had sold their house and were in the process of giving away a bunch of their things before they moved. We'd inherited their wooden lawn furniture. There were several reasons for their move, and wildfire was a big one. I hadn't yet allowed myself to process their leaving. Jay was aging, and they were headed overseas where the social contract was more securely woven. I was selfishly worried that their house would become a vacation rental. Even if it didn't, whoever moved in would be unlikely to live up to Richard and Jay's neighborly precedent; we'd become friends. Another couple we knew who had recently moved to the area was already backing away; earlier that week they left their house locked and took off to stay with family on the East Coast, with no return date. I had my own doubts too. At this point in the climate crisis it did seem, pragmatically speaking, unwise to continue to choose to live in a forest in California.

Max cut me off. I really don't want to talk about moving.

Okay, I said.

Silence.

Stay and fight? I asked.

The phrase was shorthand for us. When things got bad, a person with the choice to do so could bail out. Or you could stay and fight. When I was growing up, my parents had made jokes about running away to Canada every time the conservative president did anything stupid or scary. Around the time of the 2016 presidential election, when the fascist won, I told Max that I had always sworn to myself if the law protecting abortion rights in this country was rescinded, I would move away. I wasn't generally on board when friends joked about how horrible life in the United States was. Life here was pretty damn good. But that was my line in the sand: if I don't have bodily autonomy, I need to leave. Since that conversation so many terrible things had happened in the country and the world, and my dividing line had shifted and torn until it fell apart. In truth, soon after Max and I started dating I had already known he would never leave. He was a person dedicated to the struggle, and we both had skills that could come in handy in a potentially oppressive regime. We had an unspoken agreement: we were the stay-and-fight type of people.

It's not that, he indicated. It's just that everyone is talking about leaving and it's not . . . I don't want to talk about it anymore.

Max's reticence annoyed me. Usually, he never wanted to stop talking. I looked at him. We were both looking older lately. He had one gray hair sticking out of his ginger eyebrows. As I observed him, he became to me for a moment as inscrutable as all people are to one another, even partners, even lovers, though we never want to admit it: a person whom I could not see inside. I had no idea why he was acting this way. It felt like he wouldn't or couldn't contemplate a future life other than the one we currently had. But I could have been wrong. It wouldn't be the first time I had taken my partner's generous heart for granted. I was entirely capable of letting him take care of me without asking what he needed.

We were arguing now, except neither of us had the energy to actually argue.

I feel like . . . I began to say to Max, then stalled. I couldn't imagine a life anywhere else either. I felt as though I'd just arrived, and now I was being told to get out. Why was I listening? If I really believed this place was so connected to me, what did I owe it? To truly love a place, a person needed to take responsibility for her involvement with it, not only feel feelings. Though I couldn't imagine leaving, I could no longer guess what it might look like to stay.

In the heat, the easy hike had pushed my body past its limits. I'm having some pain, I said. Whenever I said this to Max it was usually true. However, I sometimes didn't consciously note it until the pinpricks had already ascended from the depths of my body, making me prickly too, and I had to train myself to express my pain verbally even when it might seem to me obvious. Pulling the pain card had also become an easy way for me to end a conversation or an activity—to say, Stop, I want to go home now. The two functions were not unrelated: sometimes when I felt annoyed with Max, when we argued, I would realize belatedly that I was mad already, because I was in pain.

On the way back to the car I lagged behind, the pain-hook in my core tugging my attention downward. Max went ahead and I stopped, noticing the sensations of a warm forest at sunset. The landscape around me betrayed its crisis: The trees were so close together they made a darkness of themselves. The soil was dust. The groundwater beneath it was diminishing. How dry it was. How ready to burn. It *wanted* to burn.

If it was this dry in the redwoods, I couldn't imagine what it was like east of here, in the hot hills and grassy flats. A new wildfire had started earlier that morning. First reported was a fire named Shady, near a winery in the Napa town of St. Helena, which initially flurried east, then somehow jumped the valley and headed west toward the Sonoma County line. Soon another start was reported, likely ignited by a wind-driven ember, and this second start accelerated into a second fire, and the two whirled together into what was now named the Glass Fire. As Max and I hiked, the Glass Fire was moving west through the Mayacamas Mountains. The Mayacamas were one of California's coastal ranges, and they created the border between Sonoma County to the west and Napa and Lake Counties to the east. *Mayacamas* was thought to be the Spanish transliteration of their original Wappo name. There were certain gulches in those hills that acted as funnels for wind and, accordingly, flames. In 2015, 2017, 2019, and long before that, destructive wildfires had started there. The Mayacamas were a diverse range, containing within their short span a dormant volcano, native oak woodlands, redwoods, dry chaparral, and abundant conifers. They also contained wineries, hot springs, hippie compounds, a wild-game preserve, a petrified forest, meditation retreats, wedding venues, hilltop mansions, state parks, horse farms, schools, twenty-two power plants, and an ecological reserve dedicated to studying fire resilience that itself had burned twice since 2017.

As I trudged on the watershed trail farther west, lost inside my own body, I heard a bird call or maybe it was a hiker laughing, and I looked behind me toward the vista point. From some of the trails in this park, I knew from experience, this landscape appeared pristine. No power lines were visible. No buildings, no highways. A hiker, if she didn't know better, might come upon this view and think herself to be in an untouched place. I didn't need this forest to be untouched anymore, I thought. I needed it to survive.

A FUTURE MEMORY. I'm standing in our yard near the apple trees, on the hill that is no longer overgrown. I'm with Sasha Berleman, a fire scientist, wildland firefighter, and director of Fire Forward, a nonprofit that trains volunteers, landowners, and other interested individuals to conduct prescribed burns. I'm showing her the first wildflowers of the year in these woods. The winter after Max cleared the hill in the yard, we began to see large conical leaves appearing in threes, pushing up through dirt, duff, and the gravel in the paths. The leaves had faint brown polka dots on them; within days, small flowers emerged, maroon-and-white-striped triads of petals that flopped down to the ground on long stems like strings. These are the most gothic of early wildflowers: a member of the lily family named fetid adder's-tongue, also known as slinkpod. Fetid is because they smell bad, but it's an antique sort of stink that doesn't offend. Now the eastern flank of the yard is covered in hundreds of them; I've never seen so many in one place before. The flowers are pollinated by a gnat that lives on a certain fungus. I have read that sometimes ants and slugs also aid in the plant's distribution, bursting open the seed pods when eating them, and slugs are ever present in my flower beds; the former logging camp is notoriously moist in wintertime. It's late winter now, almost spring. The smell of smoke drifts over my fence from a neighbor's yard; he's been burning a pile of yard trimmings all day.

I'm interviewing Sasha about her work with good fire, but I'm really showing her my garden and the forest behind it. Sasha is a small white woman who initially looks more like she belongs on a gymnastics team than on a hotshot fire crew. We are talking about spotted towhees, a bird that lives in the yard and that I also have tattooed on my arm, when I decide to go ahead and ask her what I really want to know.

The way I phrase it isn't elegant: We're fucked, aren't we? Like, with fires?

Game over hovers above the conversation like a fly.

Sasha looks away from me and smiles. I think she's going to say something about how if we do enough work to manage the forest, we can change the outcome, stop the fires. I don't want her to offer such a promise; I wouldn't believe her if she did. But she doesn't say that.

I don't think of it that way, she says. I mean, yeah, fires are coming, we can't stop them. But we can do our best to ensure that when fire comes here it's part of the inbuilt cycle of nature, it's doing the good work and not the bad work.

She bends down to look more closely at the wildflowers. She's excited, she's having fun. She asks if I can save her some seeds from these flowers. We can sprinkle them on burn areas, she says.

The best part of her job, she goes on to tell me, is when she teaches others how to do burns. Good fire changes fear into action, she says. Humans have an innate connection with fire, we evolved with fire, in some ways *because* of fire, and we have a relationship with it that is super primal. Rooted. Reestablishing that connection is what she sees her work as doing.

Sasha began her career in fire science, then wildland fire suppression. I've watched her TED Talk, in which she describes growing up in Southern California with fire as a looming threat on the ridgeline. She simply hadn't known that humans could have a positive relationship with fire. When she learned about the history of Indigenous burning in college, she became committed to good fire.

It's important to me, she says, to always keep acknowledging and supporting the role of Native people in good fire. Indigenous people have been doing this work for millennia, worldwide. They're still here, and they're still practicing.

And we are here, too, I say, as I steer us out of the yard and into the forest at large. Dead tan oaks teeter on either side of us.

The land that we live on is still dependent on people to steward it, Sasha says. So if we're going to be here, we have a responsibility to support Indigenous fire leadership and practices. And to take responsibility for stewarding this land as well. Not for *people,* we're fucked, I guess—she laughs—but for the actual land.

Since I moved to Sonoma County I've heard a lot of people laud the concept of stewardship. Stewardship always sounded to me like something nonprofit organizations tell volunteers they're doing, when what they're actually doing is picking up trash. It's a feel-good phrase with a malleable definition, much like conservation. I know that Sasha, as a nonprofit director, has her elevator speech dialed in. I also understand she didn't originate this concept; it's a central tenet of Indigenous and land-reliant ecological approaches everywhere. But I don't think Sasha's giving me a rap. She believes this. To me there's something wild, something daring, about simply saying that people should support the future health of the environment we live in—not for our own benefit but for that of the land and its other occupants. This definition of stewardship floats the possibility that humans are not the main characters in the great drama of Earth.

Good fire isn't a fix-all, Sasha clarifies, but it is a way to try to restore balance to the land. We have to set this land up for success. It's not actually about us.

She glances up at the trees. Her wet raincoat floats against the sea of green foliage. She looks much younger than she is, and she is far more fierce than she looks.

BY YOM KIPPUR, the morning after Max and I walked on the watershed trail, bulky kernels of ash from the Glass Fire were falling. The ash settled on our cars, the dahlias, the lonely back deck. More burned bay-leaf travelers drifted through. The pallor of unburned matter covered the abandoned traces of our backyard activities. Inside days were upon us once more.

Max and I wore white. With the exception of my morning tea—I wasn't a masochist—we fasted. I was spacey and hungry. Lydia D., a photographer friend from the city, drove up and we sat in the living room, masked, but with the windows closed to the redoubling smoke. Lydia: sunshine attitude, queer joy, artsy vibes. Lydia was also Jewish; her mother, of whom I'd once written a magazine profile, was a Holocaust survivor. Lydia had brought along a book called *This Is Real and You Are Completely Unprepared,* which was a sort of Buddhist-Jewish meditation on the Days of Awe. I liked the title. I skimmed through the first couple of chapters, in which the author, Rabbi Alan Lew, described Yom Kippur as a day of "rehearsing our own death." The holiday ritual, he wrote, occurred at this time of year because that was when people awaited the first rain after a long dry season. And as everyone in California knew, an end to the dry season was never a guarantee; one had to be prepared for death as much as life.

After Lydia left I changed into nonwhite pajamas and did my nightly routine of watching the slow apocalypse unfurl across my tiny blue screen, rapt. I rehearsed my own death all the time. Scroll, work, scroll. It was infuriating, how repetitive it had become: the fire was heading west, the fire was uncontrolled, the fire was within city limits, again. Mental inventory: go bag, keys, nonsynthetic shoes, all we had to lose, all that was never ours to begin with.

The stars outside were veiled in smokenight. I could feel the weight of the smoke, the human things and earth things it was made of, the way it retarded the breath, the pulse-ache throughout my head. I felt the unburned parts of the burned things entering my lungs. I tucked the phone under my pillow and spooned into Max, smelling the ruddy warmth of him. I closed my eyes, that nightly practice of rehearsing death that everyone did, and I thought that when it happens, if it could happen in this bed, I would be grateful. This place. With this person. The living silence of the trees. The pink-white rosebush that smells so good. Friends on beaches. Anomalous blue skies that come for hours or days at a time between smoke clouds, bringing a chill to the air that hints at humidity. My pillow glowed in the unnatural heat of the bedroom, and I realized I was living inside endless days of awe, with combusted pieces of trees and animals and people falling through the air outside, where I no longer went.

October

· 8 ·

Devils

In the first week of October the wind changed and blew the smoke to the west. The fires worked overtime to consume more fuel; the land and its residents worked in vain to withstand the heat. Rebuilding work continued on houses, businesses, and infrastructure that had burned in 2017, 2018, 2019. Carpenters framed, insurance agents adjusted. Volunteer firemen took leaves from their day jobs to go work the line on the megafires, sometimes far from home. Across the western United States, the people who worked outside worked in smoke. Air filters worked overtime in offices and homes. Water pumps worked to irrigate crops while fracking drills worked to use more water to blast open the earth's innards. Factories extended long lead times and shorted staffs. Professionals worked on video calls with their children on video calls in the background. If people had stopped working during the pandemic or the fires, they got new jobs or returned to their old ones again, as soon as possible. It was all essential work, the work of the world. If everyone worked hard enough, soon life might return to normal, to keep growing, keep profiting, chop-chop, never, ever stop. On weekdays I sat in the living room doing emails while Max paced the spare bedroom doing calls. Time accrued, as did my indignation. We had all been working, this whole time. When did we say, This is enough, no more business as usual? Our worlds were burning, and we were at work.

In Sonoma County the landscape continued its business of fire season, marching toward another drought winter. In the hills the poison oak was now a beautiful sunset-streaked fuschia. The hills themselves were gold, then brown, then a browner-than-brown noncolor. In the forest where I lived the bigleaf maple trees were turning. Soon all of the outdoors would be crowded with their large yellow leaves. Fire season was also duff season: the redwoods too were beginning to show color at their tips, dead needles like bleached highlights in their coiffures. It was a natural part of the trees' annual cycle, but this time of year I always worried that they were dying, maybe it had finally gotten too hot. The garden, too, began to ease toward its annual cycle of death and renewal. Sunflowers tilted, cosmos went to seed, and the roses slowed down. My bouquets started to contain more members of the Asteraceae family, the late-season, heat-loving perennials so beloved by pollinators and lazy gardeners. The re-meadowed yard turned from a gold carpet supporting spires of gone-to-seed wild carrot flowers into something more akin to plain dirt. In the afternoons the sun slipped lower and lower in the sky, so that when I drove west toward home it felt always as though the light was in my eyes. Despite the onset of shorter days the weather grew hotter, and as October began it felt as though I would spend the rest of my life trying to escape fires on the ground and that blazing star in the sky.

When I wasn't working or trying to write while listening to Max work, I tended the garden in an N95 mask. I'd never have the energy to rake the voluminous redwood duff into piles, but I walked around cutting back perennials that were done blooming. I deadheaded roses and brushed ash and dust off the leaves of the lemon tree. I called what I did in the garden work but it wasn't. There was pleasure in the work of tending a garden, and relaxation in reaping its rewards of useless beauty. I understood, however, the difference between horticulture and agriculture. While a home garden was a place of leisure, a farm was a place of labor.

In Sonoma and neighboring Napa Counties the main economies were agriculture, which meant wine, and tourism, which also meant wine. Whenever there were fires, which was every year lately, the national media liked to refer to them as the Wine Country Fires. Wine grapes were a luxury monocrop, having replaced apples and plums as the dominant crops in Sonoma County since the mid-twentieth century. There were wineries all over, and endless grids of clear-cut vineyards bordered most roads. What the news and the vacation rental listings called wine country was a lifestyle brand, supported by labor of many types. I lived in wine country but I also lived in waiter, housecleaner, contractor country. It was teacher and farmworker country, too. The people of Sonoma County were fixed-income retirees, small-shop owners, administrators, nonprofit workers, nurses. I didn't know any vintners.

In the wine business, fire season was also harvest season. The grape harvest began in August and often lasted through early November. Wine was an estimated $7.6 billion industry in Sonoma, a county with a population of just under half a million people. In 2020, wine grapes grown on Sonoma County soil sold for nearly double the average price of grapes grown elsewhere in California. (Only grapes from neighboring Napa County cost more.) During harvest season at night I'd see large trucks carrying plastic bins of grapes, driving from vineyards to wineries. Most grape varieties were harvested at night, the ideal time for the grapes' flavor, and growers towed stadium-style floodlights into the fields to illuminate the workers. Ninety percent of the farmworkers in Sonoma County worked exclusively in vineyards. Grape picking was piecework. People were paid by the ton, so in order to pick as much as possible they literally ran, carrying cartons of grapes to the weigh station, where they slowed down only enough to carefully pour the grapes into larger, two-ton crates. Every night of every harvest season, Sonoma's country roads were lined with parked Hondas and dented minivans, the workers' cars.

During Wine Country Fires, the labor behind wine country kept working. Vineyards were allowed to bring workers behind evacuation lines in order to harvest. At night a person driving away from their home in an evacuation zone might see the floodlights in the fields, smoke bisecting beams of light midair, workers running to bring in the harvest before the fires came closer. In the event that the grape harvest was smoke-tainted or rendered otherwise unusable by wildfire conditions, wineries could be covered by federally subsidized crop insurance, which paid out $63 million to Sonoma wineries after the fire-related crop losses of the 2020 harvest season. The people who picked the grapes had no such securities. According to the Economic Policy Institute, in 2017 most farmworkers in California made $17,500 a year on average. For farmworkers in Sonoma County—where by 2020 the median home price was $700,000—harvest season was when they made most of their money. If they didn't work, they didn't get paid.

Vineyard labor was done by a specific group of people: 97 percent of farmworkers in California were Latinx, with the overwhelming majority being immigrants; the California Institute of Labor Studies estimated that 20 percent of farmworkers in the state were also Indigenous. In Sonoma County, many farmworkers were originally from the areas of Oaxaca and Michoacán. What sent them here was, in part, climate change. Now their labor was being used to grow monocrops that sapped the soil of nutrients, which were irrigated with water the land didn't have to spare and treated with pesticides that polluted rivers and wells, in a place where the soil was branded as a luxury item and real estate prices were so outlandish that if a farmer wanted to grow food instead of liquor, they had a hell of a time finding a place to do it.

A FUTURE MEMORY. It's summer in Sebastopol, a town of seven thousand people about twenty minutes east of the former logging camp. Max and I are picking up vegetables at our CSA—community supported agriculture, in which members pay a yearly amount to receive vegetables directly from a small farm. The couple who runs it, David and Kayta, are some of the best small organic growers I've met in all my time being my dad's daughter. They're both quiet people, hard workers, so it's easy to miss the fact that they're also visionaries. Their vision is growing food for people. Lots of it. The farm is at the southern edge of the Laguna de Santa Rosa, the largest freshwater wetland in coastal California. In winter when it rains the laguna expands and half of David and Kayta's fields are underwater. They'll take flood over drought any day—the last time they farmed was that fiery summer of 2020, on different land farther west, but the following winter it didn't rain enough to fill the irrigation pond; David and Kayta couldn't run their CSA because they didn't have enough water to grow vegetables. They found this land on the laguna after that, through some older small farmers who had retired. It's 2022 now, their first season back.

I'm in the CSA flower garden, and I'm in heaven. The laguna is visible beyond the shade of several hundred-year-old oak trees. Bees buzz. The lisianthuses are as tall as I am, their watery blue cups facing the sky. It's ferociously hot out, again. This year we've so far been spared a huge wildfire, but it's early yet. As I pick flowers in Kayta's overflowing beds, Max harvests cherry tomatoes out in the fields. Unlike in the city, where a produce box gets delivered to you once a week, our CSA is very hands-on. In addition to the weekly haul of harvested vegetables, one of the perks is unlimited access to u-pick crops. This allows the farmers to grow more crops than they have labor to harvest, and it gives the customers more variety. All-you-can-cut flowers. A strawberry patch. An herb garden. Aisles and aisles of tomatoes. Bottomless bucketsful of fresh basil. The irony of u-pick crops is not lost on me: I often wisecrack to Max that a great pleasure of being middle-class is paying to play at being a farmworker, without any of the realities of living that life. I don't care, though. Each week I get to create abundant bouquets, more stems than could ever grow in my north-facing garden in the woods. If my flower garden is respite, Kayta's is paradise.

I hear Max talking to someone in Spanish; his voice cuts across the fields. He's greeting a small woman with long black hair. She wears skinny jeans and a long-sleeved shirt, work gloves, and a sequined baseball hat. David, the farmer, is with them, his pale face eternally burned ruddy despite his large straw hat. He looks tired, as all farmers do.

The woman Max is talking with is Anayeli Guzman, a farmworker he knows through his new job as a community organizer. Anayeli usually works the wine grapes. She and her colleagues don't often work at small farms; these types of places rarely have the money to bring on much help. But it's a lull right now, between pruning and harvest, and she's recently been hired part-time by David. She is weeding a watermelon patch when she spots Max's distinctive gait across the field; she walks over to say hi. I take my armful of flowers and join them beneath the ancient oaks.

The conversation is a bit cacophonous—Max is translating between Anayeli's Spanish, David's broken Spanish, and my English. Spanish isn't Anayeli's first language, either; Mixteco is. She is one of approximately six million Mexicans who speak oral Indigenous languages, many of which are in danger of dying out. Max and Anayeli know each other pretty well, and they have their own rhythm in Spanish, which Max sometimes forgets to stop and translate. I speak Spanish poorly but understand it middlingly, and I am able to piece together the conversation.

Anayeli is a leader in the movement for farmworker climate justice; I've heard her talk before about the difficulties of working in the fields during wildfires. The night of the North Bay firestorms in 2017, she had been at work harvesting wine grapes. The wind was brutal. She could see fire on a nearby ridge. Then her childcare giver called to say their neighborhood was evacuating: the caregiver was going to San Jose, about three hours away, and Anayeli needed to come get her daughter or the woman would have to take the girl with her. Anayeli left work. She would only get paid for two hours, but in that moment she didn't care about the money. She retrieved her daughter, and they drove about an hour north to a large ranch where Anayeli had worked before, where she thought it might be safer. They slept there in the field, in the back of a truck, for three nights. They had nothing to eat, nowhere to go. When I first heard Anayeli tell this story at a community event, she cried when she spoke of her fear for her daughter during the fires. Because of Max's job, I am already familiar with the exploitative nature of farm labor—poverty, dangerous conditions, abuse—and as I listen to the conversation I expect to hear Anayeli talking about her hardships, in part because they are many, and in part because that's the limited way people like me tend to define people like her: farmworker = hardship. Those poor people. But a person's life is more than its disasters; today she isn't telling that story.

Anayeli is telling Max how this farm reminds her of home. The way the mist rises off the laguna at dawn, she says, reminds her of where she grew up in Oaxaca. Max asks what the workday is usually like back home; she says in the dry season they start farming before dawn, so that they can break before it gets too hot in the afternoon. Before they start work for the day, every day, they make an offering to the land—food, soda, mescal—to give thanks and ask for a good harvest. In the rainy season, if it rains, they go into the forest to look for mushrooms.

Anayeli explains that in her community, where they've maintained Indigenous culture, they don't pay for work. If you plant corn, she says, the neighbors help harvest. Or maybe one neighbor has cows that everyone can use to cultivate the land. In Mixteco, such community-based labor is called making *tequio;* there isn't really an equivalent term in English. If a person works the land, other people come and help. It's a reciprocal blend of labor, culture, and community care.

That's similar to how small farmers do it, says David, although more in the past than now, unfortunately.

How does the work here compare to the vineyards? Max asks. It doesn't, Anayeli replies. The relationship is totally different. Here, on the small farm, we're made to feel not like workers but part of a family. While the work is important, it's more important for them how we are. Here they do many of the same practices we did at home in Oaxaca: letting parts of the land rest, adding nutrients to the soil, rotating crops. And you don't have pesticides hurting your skin, eyes, and throat. By comparison, she says, in the vineyard there's no respect for the land or the workers. The only thing that interests them is money.

Well, there's definitely no money in small farming! jokes David. It's a little awkward; the power balances among our group are complex. I imagine that David wants to get back to work—he's looking toward the walk-in refrigerator where they keep storage crops—but he's also learning from the conversation and not wanting to order anyone around in this moment. My bouquet is getting fried in the sun, and so is my skin. We should go, I say, it was so great running into y'all. Max and I gather our bags of veggies, and the farmers go back to their work, loading squashes into cold storage.

As we go, Max looks back at the fields that border the laguna. The Mayacamas Mountains in the distance frame the view. I know he's thinking about the scene that Anayeli painted for us—the mist over the land in the morning.

Walking to the car, he says, It's so ridiculous how we call farmworkers *workers* and not *farmers*. Indigenous migrants have so much knowledge about the land, he says, if anyone would listen to them. She should be running her own farm.

In early October, Max packed his travel bag for the first time since the pandemic began. He had been asked by his union to spend the month in Reno, Nevada, canvassing for the nonfascist candidate in the upcoming presidential election. He'd be door-knocking against odds in desert heat during an airborne pandemic. We had talked it over endlessly—would he go? How could it be safe? What would happen if there was a fire here while he was gone? There were fires near Reno, too. Before the pandemic Max had traveled frequently for work, and we both valued the time and independence that setup allowed us. After so much wildfire fear—and six months of pandemic isolation—I was unable to conjure a positive vision of a month spent without my person. But the rising far-right fascism in the United States was real, and Max had the skills to organize against it. He could talk to anyone, pull the life story out of them, and show them how to channel it into action. We both knew he had to go. And so he was going, driving solo for Covid safety, and I was staying home during fire season, alone. He checked with work to make sure he'd have his own hotel room, with a kitchen. Then he went to the grocery store and returned with enough food to last a month.

Before he left, we went to look at trailers. This was a stopgap of sorts regarding the question of whether or not to move away; we hadn't revived the conversation directly, although we were still being asked about it frequently by others. Since the start of the pandemic, there had been a cultural wave among mostly white, well-off people in which we all became fascinated with vehicular living. People were traveling in trailers and vans, vacationing while working remotely and without the complications of encountering other humans. This felt to me similar to the irony of the u-pick, but with much more expensive products in hand: people with money buying things that allowed them to live more like people without money. There were already a lot of people living without houses in the woods in West County, and there had been several recent small starts—fires that started but were extinguished before spreading—in an area where unhoused people were known to camp, sometimes lighting fires for cooking or warmth.

Because West County was a rural area with one main industry—tourism—there weren't a lot of full-time or year-round jobs or homes for working-class people. There were vacation houses in varied states of repair: luxury retreat homes that cost millions of dollars and uninsulated summer cabins passed down through families for generations. And there wasn't much else. Some of the available housing stock was tied up in the form of vacation rentals, which stood empty in the off days and months when they weren't making money for the tech company that by 2020 had already ruined travel and real estate all over the world. Overabundant short-term rentals meant fewer long-term rentals. Homes nobody lived in sat fallow, while people who couldn't afford the astronomical rents for the rare vacancies created ad hoc compounds of trailers, tents, and mobile homes on friends' properties, or slept rough in cars in the woods.

Wildfires had made the housing situation worse. They had destroyed affordable neighborhoods and trailer parks. High construction costs meant that rebuilt homes were more likely to be occupied by wealthy people. When wildfires burned wealthy areas (the rich tended to live high on hilltops, where fire also enjoyed ideal conditions), the displacement could have a chain effect. I had heard repeated stories of renters being evicted after their landlord's primary home burned down and the landlord needed a place to live while rebuilding. In my neighborhood, the effects of polycrisis were tangible. Max and I were ourselves examples of the crisis: We were technically rural gentrifiers, having been pushed out of the city and taken advantage of cheaper housing costs in a rural area, in the process driving up home prices and adding fuel to the WUI. Our neighbors who were leaving, Richard and Jay, were doing so in part because the economic crash of the pandemic was pushing them out of their retirement idyll. In early September, when Max and I returned from the Sierra Nevada, we had noticed a tent pitched on our road, in one of many pockets of forest that reached down into the neighborhood between house lots; a man was living inside it. Living unhoused wasn't a trend; it was a situation in which real people really suffered, and I saw it everywhere I went in California. I didn't find it enjoyable or acceptable to make a hobby out of the lifestyle aspects of being unhoused, without having to experience the accompanying risks.

I was, however, acutely aware of the inevitable fire evacuations in my future. A trailer might be a good evacuation solution: a way to leave our home temporarily, when we had to. Max and I were camping people—we had lived in a tent for six months while bike touring—but we now knew that tents weren't ideal when the air was made of smoke. The cost of a trailer ran from a few thousand dollars to many tens of thousands; we still had some savings from our apartment buyout and could afford something used and small. Over the past few weeks, I had started browsing vintage trailer forums online, trying to picture myself in a diminutive tin room, evacuated but at home in the world. First it was a lark, a way to waste time online; soon I found myself setting up appointments to see them.

The vintage trailer in the deep East Bay was shiny and chrome. It was owned by a young second-generation cop. Inside, the trailer was huge. And heavy; it didn't feel very mobile. I imagined cooking dinner in there wearing my ratty old Mother Hubbard apron, cosplaying some sort of 1950s sitcom, bomb-shelter life. Even my dark sense of humor didn't find this unlikely scenario appealing. I couldn't picture us hooking this metal box to a truck we didn't yet have, or figuring out how to turn it around on our narrow street during an evacuation. Besides, the electrical wiring inside the cupboards was visibly old, and Max, an anarcho-syndicalist, and I were both reluctant to hand over our savings to an agent of the prison-industrial complex.

In a 1970s fiberglass egg trailer that we looked at on the coast, the diner-style table and benches were too tightly spaced for my body to navigate comfortably. The egg was one of several expensive vintage campers that the seller kept parked in dry storage near the bay. She was a thin, linen-clad white woman with a fondness for decorating with sheepskin throws; she told us she taught yoga remotely from her Airstream, which was parked nearby. She'd paid what she said was a lot of money to *my guy* to renovate the trailer, but I thought *my guy* had done a sloppy job. Cabinet corners didn't meet up. The door didn't close fully. There was a ding in the fiberglass that the seller hadn't known was there until I found it. All this woman's influencer-lifestyle aspirations couldn't disguise the fact that this was basically a small box made of flammable, dent-able material, and she wanted $14,000 for it. Although maybe, actually, she said, she wasn't really sure if she wanted to sell it, it's just so cute, she might have to think about it. We left the trailer-hoarding yogi to her thoughts.

The day before Max left, the weather turned chilly and for a moment it felt like proper autumn: the gray in the sky was real clouds, not smoke, and it even drizzled for an hour or two. The fog adhered old fallen ash to the deck. The flowers in the garden again drooped with moisture, not heat, the air smelled of seasons, and for a moment I thought it was over, it would rain now, we were safe. Max and I headed into the forest behind the house to try to catch the redwoods drinking water from the air.

OUT OUR FRONT DOOR, over the dip in the gravel driveway, and across twenty feet of crumbling asphalt, beneath a sky cluttered with branches and power lines, there was a yellow fire hydrant at the bottom of a densely forested hill. To the left of the hydrant, between a California hazelnut bush and a drooping sword fern, a dirt gap was apparent in the underbrush. This was the path. Up, up, in, and through, and we surfaced onto a wider dirt track that followed the hip of the hillside: an old fire road. On both sides of the path, redwoods rendered the sky distant. Douglas firs, their bottom branches naked and dry, teetered above sickly tan oaks. Between them, lanky bay laurels searched for the light. All around, in the shallows, a tangle of bushes and saplings curtained an impenetrable understory. Today the moisture had brought out the colors in the forest. Miniature chartreuse lichens stood upright on fallen red branches. We slid up wet rock outcrops and poked under leaf piles for mushrooms. I remembered what it was like to feel a little chilly.

As we walked, Max and I debated trailers. I made judgy jokes about the trailer vendors and Max gamely played his usual role of enthusiast-convincer, talking through the prospect of becoming van-life people. Neither of us was feeling the usual buzz that accompanied the decision to do adventure. We agreed the specific trailers we had seen did not enchant. The general concept of a trailer, however, was still something that seemed worth looking into.

The path led us west. There were about fourteen backwoods miles from here to the Pacific Ocean; it could be walked, if you didn't mind a few private property signs, and nobody did. Max and I had done it on a foggy day, much mistier than today. It had been a grueling experience for my body, but I loved the route because it took us through a perfect cross section of west Sonoma County's class hierarchies—ramshackle woods, sketchy human habitats; logged-out areas dead of spirit; private roads and wealthy compounds; a state park; the cluttered porches of former hippies who'd bought their big wooden houses when they were still cheap. The WUI was a multifaceted place. The long hike to the coast ended with a climb up a trail that had been there for thousands of years, and a stunning drop to the ocean. It was a good path, this path, our path.

It began to drizzle again. Max tugged at sprigs of Scotch broom, uprooting and dropping them on the path before they could develop their garish yellow flowers and reseed. Broom was highly invasive and essentially unkillable, but he always tried. A faded sign warned us weakly from a stump: PRIVATE PROPERTY. Max began to hum a Woody Guthrie tune.

At some point he said, I feel like a trailer is supposed to make you flexible, but it's really just more crap to be responsible for.

We don't even know how to own a house, I said, let alone an additional tiny house that is also basically a whole other car. In truth, we did know how to own a house and had done so for three years now, but neither of us had ever expected to become homeowners and we were still insecure about our ability to take care of this fallible wooden box for which we had promised our futures to a rotating cast of lenders. We were both generally uncomfortable with the idea of being property owners and all the grievous historical and political culpabilities that entailed in the Americas. The fact was not lost on me that capitalist settler colonialism had provided for me an existence in which I could declaim its evils while also owning land in California.

I pointed out to Max that I was an anxious person prone to self-criticism. It might not be a great idea for me to own yet another house, tiny though it may be, to have to worry about and repair. Besides, I wanted the house I already lived in, its perfect afternoon light, the wooden counters worn thin from years of sanding, the garish flowered carpet in the bedroom, skylights beneath trees. I wanted the place that had become for me a sanctuary, a place I credited with my bodily survival—the garden our guests always called magical when they first saw it. The little white house under big red trees on stolen land. If those places felt precarious to me, then everywhere else would too.

If the little white house became uninhabitable, it was in part because I probably shouldn't have been living there in the first place. Until this area was colonized, people hadn't lived under the redwoods. "When we went into the redwoods, we did so with purpose," wrote Greg Sarris, chairman of the Federated Indians of Graton Rancheria, the sovereign nation of Coast Miwok and Southern Pomo people in Sonoma and Marin Counties. In a 2019 essay titled "The Ancient Ones," Sarris described how the redwoods were full of useful resources, including huckleberries, ferns, mushrooms, clover, and small game like rabbits. Southern Pomo and Coast Miwok people used redwood bark as building materials for their "conical houses." Some of the same adaptations that helped redwoods survive for millennia made them great building materials for residences or acorn storage; the wood was virtually insect-proof, among other attributes. Parts of the redwood could be used in a medicine for earache or to fashion dolls for children. Growing up around coast redwoods I had heard many stories, mostly from older white people, claiming that Native Americans viewed the redwood forest as cursed. There were ancestral Indigenous stories that cautioned against going into the redwoods, but they were often transmuted into stereotypes by outsiders. "Early ethnographers characterized our culture as being predicated on black magic and fear," Greg Sarris wrote. Regardless of any spirits that may inhabit the woods, the caution of locals came from experience. Before they were hunted to extinction in California, grizzly bears lived in the forest. It was easy to get lost among the tall trees without a visible horizon or sky. Some commonly used paths cut through old groves, but redwood stands weren't smart places to make one's home. They were damp, cold, and dangerous.

It was fairly recently in human history that people had lived with the assumption that their home—meaning a structure and also a sense of place—should be stationary. Stability wasn't often a naturally occurring function of the planet on which we lived. As Max and I walked through the WUI contemplating our future there, almost one hundred million people around the world were living away from their homes, having been forcibly uprooted by violence, economic brutality, or weather. Climate migration—in which people were displaced by extreme weather, natural disaster, and unlivable conditions such as drought or heat—was predicted to become the biggest crisis in human history, potentially affecting billions of people in the coming century. It had already been happening for a while: Latin American immigrants like Anayeli had made dangerous journeys to California because their land at home was no longer farmable due to heat, flooding, erosion, or economic changes linked to geopolitics. Populations were shifting north as more regions became too hot and sustenance became scarce. In the United States, natural disasters had been shown to directly contribute to widening the wealth gap between people of color and white people. In my life this phenomenon created a paradoxical situation in which those with resources were preemptively migrating away from precarious climates. Several of my white, middle-class friends had already headed north to Oregon, only to discover that heat domes and wildfires had arrived first. This was in part what Max was resisting when he said he didn't want to talk about moving away. We were white and rich enough to entertain the fantasy of excepting ourselves from climate migration, but that fantasy contrasted with our values, including a belief in what anthropologist Anna Tsing called "collaborative survival." Besides, no amount of vehicular hardware or large impulse purchases, no fiberglass walls or chrome countertop borders would ensure the continued habitability of the place where we lived. I could nurse my white guilt or do yoga in Airstreams all I wanted. The fire would still, someday, come.

We curved up a hillside, around, then up and down again. In several places the path was blocked by piles of dead branches or entire fallen trees that had succumbed to sudden oak death. Fuel. A large madrone stretched diagonally across my sight line, its crackling red bark open like wounded skin, and I ducked beneath it. My pain today was a vulture, circling. Along the path there was also detritus of the human kind: chain-saw bites in stumps, a heap of dumped woodstove ash, and green plastic ribbon tied to select trees. We detoured around the hull of an abandoned camper trailer. Its shredded aluminum siding was peeled back like the madrone's bark. On the side was spray-painted FIRE VICTIM COFFEE PARK. They'd spelled the name of the neighborhood wrong; it was the subdivision of *Coffey* Park that burned down in Santa Rosa in 2017. We wondered if the trailer's former occupants—long gone from here—were only pretending to be fire victims. But it didn't really matter; either way, someone had come to a point at which they had dragged a trailer deep into the woods on private property in order to have a place to sleep. I looked at the trailer, which always unnerved me. When I hiked on the path, I walked as far as possible around it. A hulking tin ghost of hard times. Someone—a different someone, presumably—had spray-painted DEMOLISH ME on one of its steel tow pulls.

A FUTURE MEMORY. I am getting a tour of the trouble. I'm standing next to an old Airstream trailer. The metal on the door is warm. I'm waiting to meet my tour guide, Thea Maria Carlson, an acquaintance of an acquaintance who also grew up in Santa Cruz, although we didn't know each other there. While I wait for Thea, her neighbor, Ken, is showing me his trailer.

We're on a 414-acre property in the hot dry hills of the Mayacamas. It's classic NorCal: oaky hills, endless fir ridges, with grasslands and chaparral between. The place feels remote but the city of Santa Rosa is twenty minutes downhill. Since 1974 this land has been home to a small intentional community named Monan's Rill, after the creek that runs through it. Residents call it the Rill.

The Airstream is in a dirt clearing. A few other trailers are scattered around. Next to each I can make out the charred footprint of an old house foundation. There is one remaining shedlike building, which serves as a gathering place for the community, and one house, its brown wooden exterior looking out of place; all the other wood I can see is on trees burned black. It has been eighteen months since the Glass Fire.

Thea rolls up in a golf cart with off-road wheels that emit little eddies of dust. She's a quiet, forty-year-old white woman who wears work boots, cargo pants, and a tank top. She parks and we walk down the hill to a flat area, where a plastic greenhouse newly built from a kit stands next to a fallen, charred oak tree. While we talk, Thea works on potting up some vegetable starts in the greenhouse, and I offer to help.

I first ask Thea about the people who founded the Rill, but I say it in a thoughtless way: So, Quakers lived here first?

This was the home of the Wappo people first, Thea corrects me.

I blush. She's going to think I'm clueless. But she's patient with me. The land was occupied first by the Wappo people, she explains, then by loggers and ranchers of European descent, then by Quakers who founded the Rill in the seventies.

The Rill today espouses collectivism as a solution to alienation, consumerism, and competition. Members meet regularly to make decisions pertaining to finances and the land. Decisions are made by consensus, which can take a long time. Before the fire there were thirty people living here. Now they're down to seven, plus a family of four who are renting in town until their house gets rebuilt. Thea and her partner joined the community a couple of years ago. They live in the one house that didn't happen to burn down.

Before the fire, Thea tells me, the land around the houses was full of native oaks, which provided shade and food and habitat for the critters who lived there. There were madrones here much older than those on the path near my home, and their cool skin provided relief to warm hikers who laid their hands on it. There were two ponds, upper and lower, fed by the creek. There was a garden in a flat spot that looked out over the rest of the world. On one steep slope, dry chaparral—mostly chamise—thrived. There were manzanita—low shrub-like trees with delicate orbed leaves and ruddy complexions—with trunks wider than a person. I passed some of their remains stacked by the road on my drive up. The way Thea describes it to me, it sounds like a utopia. Ninety-eight percent of Monan's Rill burned late on the night of September 28, 2020, Yom Kippur, while I was lying in bed with my phone, rehearsing my own death.

Even in its postfire state, I feel instantly attached to the Rill, in part because its topography is so emblematic of my California and in part because the concept of a back-to-the-land community is familiar to me from growing up in Santa Cruz. California is full of failed utopias. Since the late 1800s there had been several notable back-to-the-land attempts in Sonoma County. In 1884 a Unitarian minister founded the commune of Altruria just outside Santa Rosa city limits; it fell apart after two years. (The neighborhood was later developed as Fountaingrove, a wealthy exurb that burned almost completely in the 2017 Tubbs Fire, as did a historic round barn built by a former Altrurian.) More recently, in the 1970s, western Sonoma County—then a mostly conservative, rural area—had been home to the communes Morning Star Ranch and Wheeler Ranch, which were both located within a couple of miles of the former logging camp I call home. Most utopian projects I know of ultimately fell victim to the problem of capital—how to make and treat money. Nowadays few organized communes still exist, although there are many informal residential compounds in West County.

The economic and sociopolitical realities of creating a heaven on earth are complicated. My parents' community from the 1970s had survived in part because of the widespread American appropriation of yoga and its attendant spiritualism; the Land stayed solvent by running a popular yoga retreat and teacher training center. Going back to the land was, of course, something only people possessed of land could do. We all know where the land came from. As Thea has just reminded me with her quick correction, California is still in the process of being colonized.

The communards of the West Coast counterculture movement in the 1960s and 1970s were the latest in a long tradition of Euro-Americans who, aided by generational wealth and the lasting cultural imprint of Manifest Destiny, enacted upon California varied experiments of living in greater harmony with the natural world. In many ways my current life in the woods is a direct descendant of that process. At the same time the hippies were returning to the land, its original occupants were still fighting for it. As part of the 1956 Indian Relocation Act—passed around the same time my neighbor Frank's family bought a vacation cabin as a getaway from San Francisco—the U.S. government relocated thousands of Indigenous people to dense urban areas, including the San Francisco Bay Area, with the goal of cultural assimilation. In the Bay Area in the late sixties, the movements for free love and free speech existed alongside the eighteen-month occupation and reclamation of Alcatraz Island by a group of Native activists called Indians of All Tribes.

The greenhouse is getting hot, I'm flushing. As I upturn kale shoots and transfer them to larger vessels, I take care not to disturb the hair-thin roots. I'm thinking about how the ironies of the wildfire crisis have become a constant reminder to me to question the romance of my origin story, to attempt to see through the gauzy, nostalgic filter that memory applies to any person's narrative about herself. I came from countercultural communities, I had wonderful and liberation-minded parents who encouraged me to care for the world around me—*and* I am a beneficiary of all those American doctrines of discovery and destiny, enacted over centuries through acts of removal, relocation, and reorganization.

I know Thea is from the same town and generation as me; I wonder, but don't ask, whether she struggles with some of these same contradictions. Instead I remind myself that knowing the truth about where I come from doesn't make the good parts less good; it makes them more real.

I do ask Thea if she thinks of the Rill as a utopianist project. Utopia isn't a word she identifies with. We aren't trying to create something that's separate from the larger community, she says. The Quakers were realistic and levelheaded compared to a lot of other back-to-the-land people in the seventies, Thea tells me. They were nurses and teachers, and the Rill was a way to be more involved in the community and the land.

Anyway, she says, most of the utopianists I've met either don't ever come here, or they come for a visit and we never hear from them again.

To me, the vibe at the Rill feels less like a utopian-style commune than like a loosely held tenancy in common. At first impression, the people at the Rill appear to be actively trying to engage with the land and its provenance, rather than adopting cultural signifiers of a back-to-the-land lifestyle. The kale we're planting, Thea tells me, is being donated to a food sovereignty program run by Indigenous women.

We stop potting to poke around the postfire garden, which mostly consists of a handful of raised beds surrounded by a deer fence; the rest of the garden, formerly a three-acre area that was also fenced, is overgrown. There's an amorphous pile of smelted metal over toward the road. The area that used to be the orchard is a standing graveyard of dead trees. Only four fruit trees survived. The fire burned hot and stayed long in the orchard, thanks to fresh wood-chip mulch.

Past the orchard there's an open area of ground where grass has densely regrown. A few burned bundles of chicken wire spiral up from the vegetation. I can't explain why, but when I see this spot I have no desire to get any closer to it. I notice Thea avoids it too. That's where the chickens were, she says, and gestures vaguely.

Forty chicken skulls, picked one at a time from the ashes. The barn cat, still missing, the barn gone too. Three singed but living goats. One of them was never quite the same after the fire; they had to put her down.

Visiting the burn area at the Rill feels to me like going through a portal into another world. It is a parallel world, coexisting alongside the quotidian one in which people like me are living, shopping, and working a few miles downhill from here, where tree trunks are brown and not black, where people spend our days not noticing the basic infrastructures of our lives—things like septic tanks, and the fact that chickens are living creatures, and also the verdure of the vegetation intermingled with the human stuff.

From the garden I stare out at the view. It's spring, and the new growth is stunning against all the black things. It's quiet and warm here. Thea sees me take it in.

It has an easy beauty, this place, she says.

I can see why a person would want to stay, I reply.

It will never look the same as it did, she says.

Now it is a study in contrasts, a landscape mostly made of sky. The land feels muted after the fire, Thea says. But not dead. Like its vitality is underneath the surface.

Why did you decide to stay, I ask, pushing her for more. I'm still not sure what to make of the community here. Part of me thinks they're visionary; part of me thinks they're traumatized. Probably a little bit of both.

Fire convinced her.

The community had spent the years since the 2017 Tubbs and Nuns Fires—which came very close—preparing for wildfire. They had an evacuation plan, phone tree, systems for honking horns to warn one another (cellphone service is spotty up here), and hose standpipes next to each house. They had spent endless afternoons stapling screens to the undersides of their decks and the soffits in the eaves, so windblown embers couldn't get inside. They bought scanners so they could listen to first-responder chatter. They'd done some prescribed burning and were preparing to do a lot more. It didn't matter. The fire had more than one flank and the wind was chaotic.

Thea was one of the first people back in the immediate aftermath, and she spent time alone walking through the still-smoking ashes of her community. Fifteen months earlier, they'd done prescribed burning on six acres of woodland. The surrounding areas had been decimated; all the vegetation was dead. But Thea saw that on the acres that had been treated with good fire, the Glass Fire had done no damage. It didn't burn there at all. This was a best-case scenario for land treated by prescribed fire. Thea shows me a picture of the neat, clear line between the two areas: a hard border between good fire and bad. When Thea first saw that line she experienced what amounted to an epiphany. It was possible, she realized, to make fire less murderous.

The next time fire comes to this land—and it will—they'll be more ready. In addition to managing the vegetation with future conflagrations in mind, the community is rebuilding their houses from noncombustible materials; hardscaping and landscaping are being done in ways designed to keep flames from spreading between structures and vegetation. Thea says they want the houses to be survivable without anyone defending them. During the Glass Fire, firefighters never made it up the road to the Rill; megafires make resources scarce. If there had been someone here to prep the houses, Thea thinks, they might not have been a total loss. We want to make it safe enough, she says, so if anyone ended up having to stay here, they would be able to survive.

Wait, you'd stay here *during* a fire? I ask.

She isn't saying that. Some folks on a neighboring property stayed to defend their home from the Glass Fire and were injured—and they were firefighters. It's more that sometimes not everyone wants to, or can, leave. She clarifies: The idea is to make it safer to stay, it doesn't mean everyone *will* stay. We just want to make both those options as safe as possible.

I once attended a presentation by a wildlife photographer who was obsessed with mountain lions. He used trail cameras to track them and dreamed of encountering one in person. An audience member asked if he was afraid of being attacked by a cougar, and he replied no. He said, I mean, if you gotta go, that's a pretty awesome way to go. At the time I thought he was an egoist—a lone, enraptured male. But talking with Thea I begin to wonder if there isn't a refreshing sort of clarity in such a relationship to mortality. She has thought realistically about the possible outcomes of continuing to live where she lives. She has made a decision; she knows what it might mean.

I walk across the grass to the picnic bench, seeking shade, and Thea points out that I'm stepping on the footprint of the old greenhouse.

I look around, across the hills that seem never to end, thousands of burned trees still standing where they died. Do you miss the old trees? I ask.

How old is a tree? she asks back. The burned manzanita are already sprouting from their bases. They were resprouts themselves before they burned. So is a tree as old as its wood? Or is it as old as its roots, that perpetual thrust of life regenerating century after century with little regard for the micropolitics of why the fire has come, only the knowledge somewhere inside its smallest cells that after the burn it is time to start again?

It's such beautiful land, I say.

You don't see the ghosts of the houses, she says.

Do you see ghosts like that all the time? I ask. The ghosts of the old places?

Not all the time, she says.

ON THE PATH in the chilly western woods that early October evening in 2020, while the Glass Fire still burned to the east, Max and I reversed course and headed back toward the little wooden box in which we lived. Past where we'd turned around there was another pathlet that led to a small redwood canyon—fern gully, I called it—which led a couple of miles uphill to a secret swimming spot, a disused reservoir that had served as the (questionably safe) water source for the former logging camp until in the 1990s the neighborhood had contracted with a new water utility, which sourced better water from the river. It was an enchanting hike, especially when the weather wasn't apocalypse hot, but my body wanted the house.

Before we returned home I took a picture of Max standing on a steep incline on the path, atop ochre layers of duff. He was wearing orange, which was also the color of his hair, and his eyes were golden as the hills of California. He looked to me like autumn—not fire season but proper, cinematic autumn, those flaming hues of crisp afternoons, wet morning dirt, and comfort. Sweater weather. A dusty sunbeam broke through the trees and spotlighted him. Red beard, red bark, umber forest floor below. In the sunbeam, a cloud of insects became visible and gently haloed us both. Fairies, I whispered.

The next morning the heat came back and sucked us dry again. I poked around the garden and gathered Max a bon voyage bouquet, something I used to do for houseguests, back when we still had house-guests. All the flowers I found were the colors of flames—marigolds, apricot roses, and one glorious red dahlia. I added a fallen scrub-jay feather and a small branch of David and Kayta's cherry tomatoes for flair. I arranged everything in a small jar and tucked it into the drink holder of Max's car while he crammed his boxes of food into the trunk. Then he was gone, and I was alone with the wind.

It was an ill wind. It came from the northeast, beginning in the high pressure of the Great Basin and gaining speed as it descended through the foothills of the Sierra Nevada. As it rolled through the Central Valley it collected heat, marshaling stamina and pushing over the coastal ranges and through canyons, whipping across everything between Mount Diablo and the Pacific. The wind was what kept the marine layer offshore during fire season. When it blew through the redwood forest on the hill where I lived, the first hint that it was coming was a chime on my front porch. Then the wisteria leaves shimmied. Soon, a sound like the ocean but somehow drier began to relay through the canopy, first up the hill in the firs and redwoods, then down through the oaks on the path and into my backyard. The trees above the house danced with abandon; looking up at them when it was windy would spell me into dizziness. The night Max left for Reno the wind entered the house. It gathered in certain corners: by the kitchen window, in the corner of the bedroom behind my nightstand. The wind was invisible but unignorable. It liked the night best of all, and in starlight hours it would fly at fifty miles an hour through the trees, toppling some of them. On gentler afternoons it nudged the dandelion seeds across the yard.

The wind was named Diablo, the devil, katabatic cousin to Southern California's trickster Santa Ana winds. Diablo winds were named for Mount Diablo in the East Bay Area, a lone pyramid of earth and volcanic rock that rose from the flat floodplains of the Sacramento

River and stood as a beacon for those traveling inland toward the Central Valley. The mountain's devil moniker had uncertain origins but was thought to be a misnomer first applied by Spanish soldiers during colonization in the 1700s. While in violent pursual of several Bay Miwok citizens, the soldiers had become lost in the brush on the mountain's northern slope. As the story went, they nicknamed it *monte* (thicket) *diablo* (of the devil), which at some point was transliterated into Mount Diablo. The real names of the mountain were much older and not for me to know.

In the early twenty-first century, the word traumatized was overused in nonclinical settings, but it was the correct term to describe how anyone who lived in a fire-vulnerable place felt whenever the wind blew. In 2017, the wind had blown at ninety miles an hour during ninety-degree heat, turning a spark from electric equipment in Napa into the Tubbs and Nuns Fires, which together killed twenty-five people and burned approximately ninety-six thousand acres and seven thousand structures, many of them within a span of several hours. That wind, and the fire it pushed, had entered the city of Santa Rosa and burned down the entire subdivision of Coffey Park. In 2019, the wind that spread the Kincade Fire, which burned northeast of Santa Rosa near Healdsburg, was clocked at ninety-five miles an hour. Back in 1991, I had watched on television as the Diablo drove the destructive Oakland Hills fires. In a wildfire, it was winds like these that lofted embers for a mile or more ahead of a blaze, starting flare-ups that could then spread in new directions, as was currently happening in the Glass Fire; or, as had happened in the Santa Cruz Mountains in August, the wind could pivot and cause smaller flare-ups to burn backward and join up with the initial fire. If the charred bay leaves that rained in my backyard on the nights of those fires had been drier or hotter, or the wind had worked differently and guarded them like coals as they flew through the air, they would have beset the spaces between the deck planks or under the eaves of the house, glided through air vents and roofing tiles, and burned. When these winds blew, people in this region knew things were about to get very bad.

When the devil winds didn't bring fire, they brought sensory flashbacks and paralytic worry. The wind pushed its way inside my collarbones and up the nape of my neck, irritating my eyes until they wept. It made my legs and backside incongruously cold in the heat, and pain begin to prime itself inside me. The wind irritated, it overtook, it made me so damn angry. It would be a year or two before there'd be a study specifically on repeat wildfire threats and emotional health, but I didn't need academic peer review in order to know what it felt like when a switch was flipped and my lizard brain was poked. The wind undid me. Then the siren wound me back up again.

The local firehouse siren was a piercing, drawn-out noise—an immense wail that rose and fell in waves, seeming to fill every particle of the air. This fire season, it felt like a day hadn't gone by without a siren. Out here in rur-burbia, cellphone service was unreliable and municipal services were limited. The blare of the firehouse Klaxon was the traditional way to summon the Volunteer Fire Department (VFD), which served as the first responder for all non-law-enforcement emergencies, including medical calls and fires. The siren shrieked for all of them, day and night.

The siren sounded like an air-raid siren because it was one, circa World War II, although it was never used as such. Painted green to match the forest, the siren perched on the roof of the VFD, looking more like a birdhouse than a bullhorn. Its military past always seemed appropriate to me, a reminder of the general militance with which fire suppression by necessity was performed. Each sounding of the siren consisted of four keening blasts and usually lasted about a minute, during which a listener had ample time to envision the full range of catastrophes that could be presently befalling her home or her neighbors'. My neighbor Tim, who had served on the VFD for more than forty years, reassured me that most of the calls were for medical emergencies, not fires. There was an app that told me what type of event had initiated the call to 911, but it never told me what happened after that.

The siren was tremendous, ever present in daily life. It might go off several times in a day or only once in a week; its sonic reach varied with weather and landscape. It could wake me from deep sleep, stop conversations, set the neighborhood dogs to a chorus of wails. My friend Kelly once described the siren's wail as akin to the sound of a thousand raging babies, screaming throughout the forest. And at that sound, she said, as a human animal, you just jump.

These were the days of sleeping with the phone plugged in, charging, next to the bed. Then they were the days of not sleeping at all, the heat too heavy, the wind too loud, tension bearing down on my jaw until something inside my head shifted and my shoulders became granite. Every night before bed I readied the bags. One tote bag of clothes and my charged-up laptop computer, bra, and inhaler in my purse, ready to grab. Keys and mask by the door as always, headlamp next to me on the bedside table. The backpack full of our camping stuff and the actual go bag by its side. After packing the bags I tidied up the living room, as though I were leaving on a vacation and I wanted it to be neat when I came home. My response to the siren had always been anxiety: nerve synapses fired wildly, senses on alert, animal fear, fight or flight. On nights when the siren woke me with its urgent lament, I rose and stalked the windows of the house like a cat, looking for signs of smoke above the tree line. I opened the front door and smelled the night and listened for unusual sounds: bad wind, helicopter blades overhead, the monstrous roar that wildfire survivors talk about in interviews. I looked for my leather shoes. Returning to the bedroom, I put a pair of linen pants on the floor and pre-rumpled them so the legs were ready to step into. There were an estimated eleven million people living in the wildland-urban interface in California alone; I imagined each of us performing such rituals of anxiety, every night it didn't rain.

Although I complained about the siren's unnerving frequency to Max on our phone calls, over these windy October weeks alone, as my nightly prepper spiral became a muscle memory, I began to almost appreciate the siren. It was simply an aural manifestation of what was happening. At least it was direct, unlike the maelstrom of the wind or the chatter of social media fire updates. The siren said there was nothing terrifying to anticipate anymore: it had already happened.

If my response to the previous two months of near-nonstop emergency mindset had been to feel agitated every time I was afraid, then my current waking state might be described as a sense of removal that approached numbness. Somewhere beneath the rise and fall of my adrenal glands I could tell I was getting used to this. I couldn't say if that was good or bad. As the siren wailed over the treetops and the wind shook the power lines, I still inventoried my mental checklists, but the actions I took in response to alarm began to feel more like reflexes than responses.

In the daytime I did my job. I fed myself, but poorly. I emceed the monthly neighborhood fire-safety meeting without Max, who was usually my extrovert motivator when it came to public events. I tended the garden as the season turned it brown. The duff on the ground turned to dirt, and the wind blew it around. I worked on my writing in spurts, still trying to code my disaster anxiety into my characters' realities, but something about it felt rote, obligatory: I was performing the tasks of my life, not living. There were more small starts: Pack the car. Unpack. Pause the meeting. Log out, log back in. Text each other until your worries dot the daytime sky.

There were so many fires that didn't become fires. A house fire, electric, a block away. A spark at a freeway on-ramp. A utility wire rigged to a dead tree. Every time I unpacked after a false alarm, it took a little longer to reenter my life. As I tried to work, write, and live, the thud of resignation pulsed steadily in the back of my brain, the tiny Zeus of my pain now birthing a malignant Athena, waiting to bust out and lay waste to everything I had ever tried to hold together. I didn't write love letters to Max. Monotony crept in. I lost hours. More alarms sounded. I let the thousand babies cry it out.

One afternoon during the slow early weeks of Max's absence I was out in the garden secretly watering the roses. AQI 150, could be worse. We had an irrigation system set up but we tried to use it only once a week. With all these heat waves I was convinced the roses looked sad, though I understood that sadness was not a quality attributable to the genus *Rosa,* or any other plant. They pouted, they sighed in the wind. Wan ladies. They needed more. I unwound the hose from its crook and gave it to them. Max was still in Nevada, so technically I didn't have anyone to keep this secret from; he didn't know I did this somewhat regularly. I used a gentle hose setting and aimed at the soil around the perimeter of the bushes, trying not to wet their leaves, which might encourage fungus. I swore they looked happier after I gave them their secret drink. I thought they sensed my attention, my care. I knew this was a very human thing to imagine, and I also knew the roses bloomed fine with their more modest irrigation rations. But still I was here, under the smoky sky, smuggling water to them like they were stowaways.

My father, the gardening teacher, was more practiced than I was at resisting the urge to assign human emotion to plants. When he wanted to deploy a metaphor or anthropomorphize an apple tree, he'd point it out first, to be clear. He reminded his apprentices frequently that plant growth was the result of biological processes, not feelings. Plants were not *sad* when they experienced water stress or *happy* when they underwent photosynthesis. They did not *want* to be fertilized or to be placed in certain sites. They may have evolved to thrive in those conditions, but desire had little to do with it. A person cannot know what, or if, a plant feels. Yet even my father wouldn't deny he grew feelings as much as he grew plants. It was a result of science every time a flower bloomed or a fruit formed. And it was also a miracle. Gardening was an unquantifiable combination of practice (craft, rigor) and magic (chance, hope). The gardener's job was to apply science and attention. The plant's job was to act and react, according to its biological imperatives and its environment. But only one of us used metaphor to describe the experience.

Despite knowing better, having many years ago read Susan Sontag's caution against superimposing meaning or morality onto illness, I still sought in the natural world narratives that reflected my body's urgent perspective on it. I looked for metaphorical equivalences to my situation in gardening books, seed catalogs, and studies about the healing power of nature. I was always disappointed. The reproductive imperative hijacked them all. Gardening, as a domesticized symbol of nature, was often written about using a vocabulary of fertility, which I couldn't abide. Cycles of birth and rebirth, vigor and decay; it was too easy. Because the reproductive imperative was encoded in most, if not all, human origin myths, which themselves stemmed from the symbology of nature, the metaphors of the garden tended to mirror those of reproduction: Growing a garden is like raising a child. Soil is always fertile or fecund. New life springs forth from every seed, as did Eve from Adam. That was the story, at least. The story of a garden called Eden was the first story many humans learned; by some translations, the word paradise was thought to have meant orchard in ancient times. And in the orchard of paradise, morality and meaning always lurked underfoot. The garden came before the womb, but I was interested in the possibility that they might be disentangled. Just as my body was not an apple tree, my plants were not children.

I supposed it was only natural that people would take meaning from the cycles of plant life; everything on the planet was nurtured by plants. Despite the New Age vibe it had acquired in some circles, the idea of Mother Earth was a foundational spiritual and cultural concept of our species. The vision of the planet as a creator and nurturer was a powerful one. And it was a literal truth. But the personification—nature as mom—never hit quite right with me. For obvious reasons I felt affronted when presented with fertile nature. I did not believe that to give life, to nurture, to be *natural,* it was necessary to reproduce. My main problem with Mother Nature was that only half of it included me. Harmful analogies of nature as a woman—whether fertile Madonna, innocent virgin, or moneymaking whore—were as old as human nature. They had collided with modern science and medicine beginning in sixteenth-century England thanks in part to Francis Bacon, the progenitor of the scientific method. In her 1980 book *The Death of Nature: Women, Ecology, and the Scientific Revolution,* feminist scholar Carolyn Merchant wrote that Bacon's view of nature as the wellspring of valuable scientific information helped further a "transformation of the earth as a nurturing mother and womb of life into a source of secrets to be extracted for economic advance." Among other charming qualities, Bacon was fond of using metaphors of sexualized violence to describe his work in the natural sciences—language that Merchant believed was influenced by the concurrent wave of witch trials throughout Europe.

I returned my hose to its coil on the side of the house and switched to a longer one, to reach the flowers farther down the hill. The undertow inside me was strong today, and I caught myself slipping into frustration, that keen annoyance with everything and anything that signaled the quickening pulse of pain. My pelvic muscles contracted, tighter and tighter, and I sensed they were uninterested in letting go anytime soon. When I felt like this it could be days or weeks before things mellowed. People called such a rekindling of pain, or a period of more intense reactions, a flare-up. I did not find it amusing that the same word was used to describe a similar phenomenon in wildfire.

It had been announced that week that the August Lightning Complex, which was still burning in the deep forests of Mendocino County to the north, had surpassed an area of one million acres, making it the first-ever *gigafire.* The size of wildfires was obsessively measured, like sports statistics. Each new fire seemed to set some sort of record, but there was still no universal way to evaluate their damage. How did one measure wildfire? Rate of spread, acres contained, number of aircraft summoned. Structures burned, acres burned, estimated economic losses. Human deaths; although never animal deaths. Tree deaths; but the press only focused on the marquee names, like the Sequoias in Sequoia National Forest that had to be wrapped in fireproof tinfoil to avoid being burned, year after year. We were always being told how many structures a fire had destroyed, but the rest was impossible to quantify. There was no scale with which to express what fire did, how it moved, the lives it touched, the ecosystems it interacted with. It was equally inconceivable to try to measure the climate crisis, that great fire of humanity's making, fueled by fossilized trees—mycological orphans, sub-subsoil—the core muscles of this infinitely nurturing and cruel planet being sucked and drilled and blasted from beneath us, and burned. Quantifying fire was as futile as quantifying pain.

The psychologist Elaine Scarry once described pain as being both immeasurable and irrefutable. It defied description. This was in part because, Scarry said, as a bodily experience pain had no physical object, no referent in the material world. Other physical sensations had objects: You were afraid *of something*, you were hungry *for something*, you desired ____. In my experience, chronic pain was simply pain: itself and nothing else. I was in it. It had no resolution to offer, no timeline or parable that might justify its existence.

Pain couldn't even be expressed in terms of its relative severity. The pain scale—the medical standard by which patients are asked to rate their pain on a scale from zero (*no pain*) to ten (*the worst possible pain*) was by its nature subjective. Nobody could truly know how much pain another person was feeling, let alone articulate the qualities of that pain. Pain was graded on a curve. Was it still pain, I sometimes wondered when trying to value my sensations for a healthcare practitioner, if my body was so used to feeling this way that it had normalized a near-constant state of being, say, a five? Wouldn't that shift my number down to a one or a two, the values of a whiner, a psychosomatic problem? A psychologist once told me apologetically that pain was, in fact, all in my head, although that didn't mean it wasn't real. The writer Eula Biss, in an essay on her own chronic pain, had compared the ineffability of the pain scale to that of the scale used to measure wind (from *sea like a mirror* to *devastation occurs*). In *The*

Undying, her chronicle of undergoing treatment for aggressive breast cancer, poet Anne Boyer had half joked that people should replace the pain-scale descriptions with lines from Emily Dickinson poems: from *so utter* to *an element of blank*. I loved this idea; it was to me more accurate than the going modality.

I tugged the hose farther, reaching for the birdbaths. A pair of scrub jays took baths every afternoon, one at a time, the other watching as though keeping guard. One of them had a hurt wing; I called him Lefty. In the spring they nested; one year the nest fell from the redwood tree they'd built it in. I found it, put it back up on a low branch. Lefty quickly repurposed the materials and rebuilt in a nearby loquat tree, but I never saw or heard a fledgling.

The garden hose clogged, tangled, and I whipped it up and down to try to unkink its rubber layers. The effort made me double over. My pain had become acidic; it felt as though the protective barrier of my skin had dissolved. I rested for a moment against a lawn chair, then trudged back up the path slowly to untangle the hose at its point of origin. Every step was a scratch inside.

A MEMORY. Max and I are on speakerphone with a lawyer who is the mother of a friend of my brother-in-law. After my surgeries everybody I told about it said I should sue somebody for what happened to me. Max and I aren't sure who might be held responsible for my injuries, but mistakes were made. Diagnoses had been missed. A device had malfunctioned. Doctors, administrators, and pharmaceutical device safety testers had made choices. We are lucky to have health insurance through Max's work, courtesy of a marriage ceremony at San Francisco City Hall the day before my first surgery. So I hadn't incurred any debt. But companies had profited off my body and had hurt me, and perhaps it makes sense to see if I might be owed a cut.

The lawyer listens to me tell the story of what happened then tells me regretfully that the giant HMO of which I am a member is literally un-sue-able; the terms of being a member obligate members to arbitration. Besides, even if you could win in arbitration, she says, you could only get a maximum of $250,000 in the state of California for malpractice. And you'd end up giving it all to the lawyer.

Max and I are silent.

Maybe if you'd been actively trying to conceive at the time the device malfunctioned, the lawyer offers, we could sue for loss of life, a potential baby. But your uterus on its own, empty, isn't likely to win any damages. After all, your pain was only emotional. Legally speaking.

But my pain is physical, I say. Parts of my body were removed. As I speak, I am sitting with my legs curled beneath me on the couch, my body tilted to one side. A murky incessant throb emanates from somewhere adjacent to my vaginal canal. In my journal this morning I wrote, *Today I don't want to have a body; today I want to be a piece of glass.* See-through and sharp.

The lawyer is not unsympathetic but she is clear in her assessment of the language of the law. Unless someone died, she says, it's really hard to win any sort of medical case in California. Plus, she adds, you signed a waiver when you got the IUD, right?

I did sign a waiver many years earlier upon IUD insertion, while my feet were in gynecological stirrups. I've recently retrieved a copy of it from the clinic. It says I acknowledge the device *may require surgery to remove.* Just surgery: nothing more specific. Nothing about what might need to be removed along with the device. What might go wrong along the way.

Yeah, she says, maybe you could try one of those law firms that's specializing in class action suits against IUD companies? But I think they're mostly focused on the hormonal ones these days. The copper ones are considered pretty safe.

I thank her for her time and hang up.

AFTER THAT, Max and I thought about getting more referrals to more lawyers. I figured I could spend years in stuffy legal offices telling these gnarly details over and over, presenting my pain for judgment, having to prove—and believe, really sell it—that this ruined my life. Or I could write more books, plant more trees. What did I want the story of my body to be? That it was without value unless it was being used as an incubator?

Once, during a particularly bad pain flare, my therapist told me the Buddhist parable of the two arrows. Imagine being shot by an arrow, she explained. The first arrow was the cause of physical injury: it hurt. But the second arrow could hurt worse, because it wasn't an object, it was a story. The first arrow hurt; the second anticipated more hurt and crafted a scary story from the experience of that first hurt. The story, not the initial injury, was what turned pain into suffering. And it was self-inflicted. As I understood it, my therapist was suggesting that I couldn't control what had happened to me, but I could control my response, or at least the parts of my response that were under my control. The saying went *pain is inevitable, suffering is optional.* What had happened to me was not inevitable. But it had happened. So, my therapist said, what's the story you want to tell yourself about it?

I had already spent so much energy on the story of my pain. I had struggled to explain it, to express its qualities. I commenced my initial experiences of injury in a sort of beginner's mind already: as a child I had a high tolerance for pain, which as an adult I learned was in fact a high tolerance for suffering; I was highly sensitive to pain itself. When the pelvic pain began, I had to learn to identify the unprecedented sensations I was feeling as *pain,* to call them that, then to define them in terminology that telegraphed *this is really bad!* to those charged with caring for me. According to dozens of post-visit notes I read in the online interface of my HMO, at many of my most harrowing medical appointments, my condition was assessed with the phrase *the patient appears calm, in no distress.* I learned the hard way to always inflate my number on the pain scale by one or two points, because I was a person who naturally downplayed my own discomfort—partially because I prided myself on being tough and partially because so many things made me physically uncomfortable that it would not have been practical to react to all of them, all the time. Now I was being asked to undo that hard-learned vocabulary. To return to the unadorned sensations of what was happening in this body, instead of what it meant or what I could do about it. Since that first surgeon had hurt me, then paused, then hurt me again, *da capo,* I had understood that pain was not simply a matter of physical harm. Real pain was fear of pain happening again, the anticipation, the memory. The story. This was why torture existed. It wasn't about pain, it was about fear: the second arrow. To dodge that arrow, I felt I was being told, I needed to change my story. It didn't feel fair to ask this of me.

I did not want children, a fact that I had over time grown obstinately unwilling to emphasize when I told the story of what happened to me. If it had been an embryo instead of a piece of plastic and copper that attached itself to my uterine wall, I would have sought to have it medically removed all the same. In the initial aftermath after my surgeries, I used to present this fact as a counterweight to the intensity of my story, as though my lack of explicit intention to use my uterus for that particular purpose made it somehow less terrible that it had been taken out of me. I would say, I had a hysterectomy . . . but it's okay, I mean, I wasn't planning on having children. The apology instinct is strong in women and in those whose bodily experience upsets the norms of social conversation; the urge to comfort those who could or should be offering you comfort never abates. It felt like an imposition, to require friends and colleagues to sit with my experience. I managed their emotions by tacking on that relatively happy ending: All these horrors were visited upon my reproductive organs, but hey, good news, at least I didn't want kids! It was true. At least, I didn't want kids. At most, I wanted it to be my choice.

The fact of getting an IUD, the facts of the things that happened after and because of that act, were tied to my perceived uselessness as a woman who did not reproduce. My body was neither pure nor productive. Contrary to the wishes of men like Francis Bacon, there would be no secrets and no life force extracted from it. From conception, the American healthcare industry had two complementary goals: reproductive control and profit. In 1973, the year *Roe v. Wade* made abortion legal in the United States, journalists Barbara Ehrenreich and Deirdre English circulated an influential pamphlet called *Witches, Midwives, and Nurses: A History of Women Healers*, in which they traced the creation of the for-profit healthcare industry to the regulation of reproductive healthcare and the criminalization of women's medical labor. Until the mid- to late 1880s, reproductive healthcare, including birth control and abortion, was legal and widely available in the United States. Before then, reproductive healthcare had commonly been provided by midwives or traditional healers, who were overwhelmingly women. They didn't charge much for their services. Doctoring was not a profession associated with wealth; until the turn of the century, medical school was unstandardized and inexpensive to attend. Women and people of color were increasingly enrolling. That changed when the American Medical Association (AMA)—an intentionally all-male, all-white organization—was founded in 1847 and began to monetize healthcare. Outlawing abortion was a primary goal of the AMA. White doctors had noted a decline in the birth rates of middle-class white Protestant women, a dereliction of gender-essential duties that was blamed on abortion and birth control access. In 1873, using arguments that remained the basis of anti–reproductive rights campaigns through the 2022 repeal of *Roe v. Wade,* the Comstock Act severely restricted access to birth control in the United States. By 1910, due in part to lobbying by the AMA, abortion was outlawed.

With new laws about what women were allowed to do to our bodies, and who was allowed to do it, licensed doctors—95 percent of whom were men—were free to profit without competition. Women healers who stayed in the profession were demoted to nurses, while midwives and traditional healers were branded as outlaws, a scenario that again summoned the historical specter of witches—women who were punished for their medical knowledge and economic agency. The pharmaceutical industry was even more blatantly attuned to the bottom line. Within the value systems these industries manifested—and contrary to the values of millions of care workers, most of whom went into medicine in order to help people heal—both women's bodies and natural bodies were viewed as sites of extraction. Both were useless if they didn't serve the constantly expanding needs of capital.

As I watered the roses, I looked around at the vegetation. In Max's absence the weeds had begun to take back the balance of the yard. Sweet pea and ivy reached up the trunk of one of the redwoods, and I knew I should pull them off, but I was unable to make my body do any more work today. It was infuriating to me, not being able to be as productive as I wanted to be. It was a betrayal for which, I decided as

I gathered my tools to put away, I would never forgive capitalism or my body.

My pain was a map of all sorts of hidden relationships. Pain is telling you a story about your body, a different therapist had later said in a different session. Frustrated, I had thought, Which is it, then? A story I need to change, or a story I should listen to? She was simply expressing a commonsense wisdom about physical pain, which was that if something hurt, something was probably wrong. But when my pain became chronic, it stopped delivering new information. There was no discernible ebb and flow to my pain, except for the obvious triggers (stress, physical exertion). It didn't have a pattern. My day planners were filled with cryptic symbols, indexes and keys I invented and then forgot, arrows and abbreviations indicating when I thought the pain was worse and what I sensed its qualities to be. Eventually, I stopped tracking because it was depressing; the scale kept shifting, there was no arc to chart. The story of my pain was a bad story because it wasn't a story at all. As a vehicle for metaphor, pain had nothing on a garden; as a narrative, my body was a mess.

If my injury—for it was not an illness, I was not sick but rather the subject of many accidents, some systemic and some specific—had any narrative power, it took the shape of a diagram. I saw it not as an arc but as a set of concentric circles—a hurt body inside a thriving garden in the middle of a neglected forest, on an overheating planet surrounded by an unknowable universe. The story I had developed to express this experience of interconnectivity was the well-trod narrative that nature had healed me, and it was true. But that wasn't the whole story, any more than the story of fire was about man versus nature. I supposed I would always assume some inherent allegorical power in the relationship between people and plants, a resurgence of trust in cycles of growth and diminishment. It was only human to seek meaning, and meaning was often found. But the idea that there was growth to be had from injury did not always prove out. A woman was not a tree; I did not come back from my experience of being pruned more vigorous, fecund, or renewed. I had not borne fruit, wildly or otherwise. At some point, the metaphor breaks.

In a 2018 paper proposing a pyrosexual and queer ecological

theory of fire, scholars Nigel Clark and Kathryn Yusoff discussed the fetish for productivity that characterized Western ecological perspectives. Fire was perceived by dominant cultures and economies as counter-reproductive—it was a killer, not a mother. Building off the queer ecological theories of Catriona Mortimer-Sandilands, the authors proposed that uncontained fire had much in common with queer sexualities. Both were thought to be unnatural, against nature. Like nonnormative desire and love, fire was combustive, carrying in its body all the capacity to explode the status quo that that word implied. (As various Indigenous and non-Western ecologies understood, fire was also generative, not only in its ecological roles but as a force that, Clark believed, had since the dawn of humanity actively seduced humans into becoming its handlers.) The reproductive imperative in the natural world, like the destructiveness of fire, was not something that could be taken as a given.

Finished with my secret gardening, I left the hose uncoiled on the ground. I'd get it later. I sat in a chair and tried to breathe into my pelvic floor. Birds returned to the birdbath to splash. There was a threesome of mourning doves—two males and one female—that often hung out near the water. Mourning doves generally mated for life; these three appeared to be mates, and they didn't appear to be nesting or reproducing at all. When I looked it up, I learned throuples were not uncommon among mourning doves. Every evening around five P.M. they alighted on the ground near where I was sitting and hung out around the water, taking turns bathing and basking in the warmth of the golden hour. They were, like all mourning doves, skittish; the fall of a seed pod from a tree or the cackle of a woodpecker could send them fluttering up to the tops of the redwoods. Those same bossy acorn woodpeckers, whose territory encompassed our yard as well as that of many of our neighbors, were in practice polyamorous. They raised the community's young the way they did everything else: collectively. Every bird had a role, but not everyone was a parent. If I wanted to find metaphorical power in this land, there were better options than the language of fertility: the phenomenon called nature was nonreproductive, queer, and *counter* all over the place. On the paths in the forest behind the house on the hill, a small orchid that looked like a branch of coral grew in late winter. It had a haunting, almost subterranean ruddy coloring. Coralroot, or *Corallorhiza,* obtained its nutrients from fungi under the surface of the soil. It didn't photosynthesize. It was a plant that declined to use the fundamental biological function of plants. It was never green; instead, it was the color of flames.

While I watched the birds in the garden demonstrate a range of identities, to my east the Glass Fire, one week old and still spreading, delivered combustion and death to uncountable communities of organisms. In the northern part of the state, in the burn scars from the lightning fires two months earlier, the process of resurrection was already underway for many species. The coming spring would unearth a flourishing of wildflowers, many of them members of a category of flora loosely referred to as fire followers. Fire-following plants were not merely well adapted to the soil conditions, heat, and nutrients that accompanied a wildfire; they flourished in them. Some were fire-necessary or fire-obligate species, which cannot reproduce without those extreme conditions. Their seed pods or cones might lie for decades waiting to be activated by fire-strength heat or the precise chemical makeup of wildfire smoke. Some bushmallow seeds had been shown to lie dormant for one hundred years between fires. More common were fire-advantageous or fire-preferred species—plants that thrived in, but didn't require, fire to reproduce. After fire, a wildflower that had previously grown sparsely in one place might explode in volume in the same location, or appear in entirely new ground.

When it came to flowers, the seed bank's vaults were deep and varied in their inventory. In many of the places burned by the lightning fires, entire hillsides would soon be covered by a narrow-petaled purple flower called California brodiaea, which joined regulars like California poppy and lupine by the thousands. Clusters of petite but hardy yellow flowers called whispering bells would soon resound across greening slopes in abundance. To the east of Santa Rosa, on hot and often-burned Hood Mountain, fire would engender new outcrops of *Ceanothus sonomensis,* a rare red-flowered version of California lilac that was endemic to Sonoma County. In Butte County, motley two-toned harlequin lupines would come on strongly, and the minuscule flowers of short-petaled campion, a perennial herb, would appear to those who looked closely as they hiked burned foothills. In meadows between burned redwood groves at Big Basin Redwoods State Park, a tall rose-violet mallow that looked like a hollyhock, *Iliamna bakeri,* would be spotted; *Calandrinia breweri,* a relative of the common wildflower redmaids, would grow in abundance not seen in living memory. In the Austin Creek State Recreation Area, near where I lived, in springtime the shadier slopes would experience a noticeable increase of yellow three-petaled flowers called Diogenes' lanterns, a particular favorite of my stepmother's.

After the fires of 2020 in an area east of Mount Diablo, four citizen scientists out surveying native plants came across a rare fire-necessary wildflower in an area that had burned severely. After a scramble up a thousand-foot incline that before the fire was inaccessible because of its dense vegetation, the surveyors came across something none of them had seen before. There, in the hot sun, emerging from burned soil between blackened trunks of manzanita bushes, they spotted a flicker of orange. Soon, another. Suddenly, flowers. The SCU Complex Fire had been hot and fast; there was little vegetation left, even in springtime. Now, postfire, this off-trail patch of scorched land was populated by a cache of small, delicate asterisks that shone tangerine in sunlight: *Papaver californicum,* the fire poppy.

Fire poppies were the most ephemeral of fire followers, born only from the conditions that a wildfire engendered. Like those of other poppy varieties, the stems of the fire poppy were bright green and slightly hairy; they grew bending upward in a way that sometimes implied a wobble. Smaller and more graceful than the common California poppy, fire poppies grew in wider-set bunches that allowed more space for individual stems. The fire poppy's orange was a gentle one, and its petals appeared to lean toward the sun's light. Backgrounded by the charred blacks and new greens of a postfire landscape, the effect was one of radiance and grace.

October reached its midpoint and the firehouse siren kept blaring. I grew increasingly fascinated with fire-following flowers, especially the fire poppy. I looked up sightings on crowd-sourced naturalist websites and downloaded pictures taken by citizen scientists. I wondered if the fire poppy might ever grow on the hill where I lived, if it had been here before and was underfoot now, hiding since the last time it burned, before colonization. (It was highly unlikely; the environment in the redwoods was still too cold and moist, for now.) Despite my ambivalence about nature metaphors, I couldn't deny the power that the fire poppy, with its petals that flickered like flames, had over my imagination. Such beauty rising from a burn scar made me feel optimistic in a way that reproductive metaphors of the garden could not. In those first weeks of October, from my place deep within a state of constant alarm, the knowledge of these plants' existence allowed me to envision postfire landscapes that offered more than soils decimated by heat and forests made of ash. Admittedly, fire followers were a clear demonstration of metaphors of birth and rebirth in the natural world. More important to me, the flowers told a story of fire that allowed for the complexities involved in such an experience without offering up a simplified binary of reproduction or death. Unlike a woman or a gardener, fire could have it all: it was a killer and a caregiver; it played the part of villain, protagonist, and deus ex machina at once. The fire poppy photo I kept as my computer screensaver helped reassure me that my ambivalence about the cycles of harm and renewal I was experiencing was not an aberration. I could fear fire and also acknowledge the need for its return. I could shy away from anthropomorphizing plants while at the same time engaging them in what could only be called a relationship. I might refuse the reproductive imperative, but I could still interpret some confirmation of vitality in the coming of spring.

It was unlikely I would encounter a fire poppy in person unless I made it a mission, tracking sightings after fires and traveling to them before the blooms waned. Fire poppies weren't seen as often in Northern California as they were in Southern California—Napa and Sonoma were considered the northernmost part of their range. But after the 2017 fires here, some had grown nearby in Napa County. Old-timers recalled the poppies growing in the same spots after the 1964 Hanley Fire, which burned along a similar path as the 2017 conflagrations and also overlapped with some of the Glass Fire's burn areas. There were no written records of wildfires prior to colonization; Indigenous nations had other, more time-tested ways of keeping their histories. Nobody knew how long the flowers had been growing there, how many prior fires they'd lain dormant through, or in what ways they had captivated the imaginations of the different generations of people who witnessed their bloom. The truly mind-blowing thing about fire flowers was not that they thrived in places that looked like apocalypse film settings or that they lent symbolic complexity to wildfire narratives. What amazed me was the fact that the presence of a fire-following plant in a location by definition meant fire had been there before. Like the rings of Aldo Leopold's felled oak tree, the presence of fire followers showed how the history of the land is written on its body.

That next spring near Mount Diablo, namesake of the fiery autumn wind, the evanescent irruption of fire poppies would be dying by the time the hikers reached them. They only bloomed the first spring after a fire, and only once. After they faded and wilted, other growth soon replaced them, and the poppy seeds would lie dormant where they'd fallen, buried by generations of decomposed matter, waiting for the next fire to invite them to be born again.

· 9 ·

Gods

In those middling days of October, still in the depths of early Covid isolation in the United States, I had somewhat improbably made a new friend. Kelly was a white woman in her forties like me; like me, she was a writer; her body was decorated with tattoos of flora and fauna, as was mine. Her cabin looked out on a different part of the same forest that bordered my backyard, down a deader-end street than mine and beneath an even more tangled part of the woods. Kelly and I had become close by being fire buddies, and more so since Max's absence: we texted each other when the siren rang, sharing local updates. I barely knew what her face looked like, but fire had already turned us from neighbors into kin.

To get to Kelly's house from mine was a fifteen-minute walk on the road or a ten-minute walk on the path. This evening I took the path, cutting back down a secondary path that ended at Kelly's place. I arrived at her tilting red gate and pushed its rusted spring open. It was twilight. Kelly loved autumn, the turn to seasonal darkness, and Halloween; she and her eleven-year-old daughter had decorated their yard and porch for the season with spooky animal figurines. In Kelly's yard, rangy roses and volunteer hazelnuts climbed above dank soil that was habitat for real scorpions and tarantulas.

Kelly was cool, West County cool. She drove a vintage red Mercedes and her house was full of nature books and the bones of forest critters. I, too, had a small collection of bones, a little row of what I thought were skunk or raccoon skulls that I'd found in the woods and arranged on the front porch railing. But Kelly was downright witchy. She was a birth care worker. Her poems were filled with the sensual entwinement of women, animals, and plants. Lost and hurt creatures were drawn to her; she once draped a dead baby fox, found in the road, with garlands of flowers and gave it a memorial. I may have lived in and loved the woods, but Kelly was a woman *of* the woods.

She served me tea on the deck, something herbal taken from the yard. In the blue dusk baby ravens cawed from a nearby nest somewhere in the eaves of the redwood cathedral.

We traded neighborhood gossip, theories about the cause of all the small fire starts, rumors of nearby crimes. The unsettled vibe that had ushered in the fire season had grown creepier recently.

I drank my tea and my elbow bumped against a skeleton (artificial) of a small dog or wolf—part of the porch's Halloween trimmings. My movement roused Kelly's dog, Nootka, a husky-malamute mix who was napping nearby under a wooden chair. He looked around and howled at something out in the dark trees. The woods were extra spooky this evening.

A forest was a place of oxygen and life, and it was also a place where mysterious things happened. The most famous person I knew of who lived in West County was the whiskey-voiced musician Tom Waits. While this area was home to multimillion-dollar hideaways, it was also populated with people who could have been characters in his songs: folks down on their luck, maybe with drug problems and definitely with some romantic problems, sketchy backyard sheds full of who knows what. (I knew several people who knew where, exactly, Waits lived and would never tell, which was another thing I loved about West County.) People took to the woods of West County to hide. To be left alone. Sometimes to do bad things. As a popular subject of legends and lore, the uncanny forest had a way of reeling people in. The forest could heal you and shelter you. It could trick you, it could ensnare you or twist you up or starve you, if it wanted to. And if it was disrespected, or if it was injured, the forest would kill.

It had been hot out again, but this evening delivered a suggestion of autumn chill, and Kelly was wearing a celebratory beanie and hoodie. Her eyes were the color of fog. She showed me an article about living with fire in which she'd been interviewed about the experience of evacuating with her daughter. We laughed about the illustration of her, which rendered her blond (she was not) and, as she said, basic. We talked about the weather in that way of burdened meaning that all weather chat had taken on as climate change accelerated. Kelly told me that in the springtime she had noticed an unusual number of juvenile critters out and about far earlier than usual—birds out of nests, young foxes wandering far from their dens. Her theory was that early heat waves made the animals' dens too hot to stay inside. As we talked, an acorn woodpecker squawked. A bird siren. Acorn woodpeckers were killers sometimes, too. I'd recently heard a mob of them nearby, congregating in one tree and making vocalizations that I could only describe as yelling. It went on for hours. In addition to raising their young collectively, acorn woodpeckers had a complicated hierarchy of mating positions. They mated for life. When a mating male woodpecker died, there was a fight—a brawl, really—among the collective to determine which birds (usually a set of siblings) would be promoted to the status of breeder. These mob fights could last for days at a time.

On the way home from Kelly's I took the streets and not the path; it was dark now, and the woods contained that familiar threat of unseen things lurking in shadows. As I lay in bed next to my keys and headlamp, the moon shone through the scrim of redwood needles and onto my face. I must have slept because soon I woke to a loud wail, afraid. Assuming it was the siren, I looked to the window for smoke. I reached for my phone, but there was no notification. The noise continued, and I realized that it was the cries of a pair of foxes. Although I knew these were not the type of screeching sounds that required my action, my body wouldn't stand down. I dropped my phone and pawed the bedside table for my headlamp; the foxes babbled their foxy song. I counted my breaths and tried to visualize a collective forcefield of vigilance—my neighbors, the critters, the redwood trees more powerful than all of us—surrounding me, keeping watch. I closed my eyes and listened as the foxes bleated their alarm into the bosky night. Wake up, they seemed to be crying. This is really happening.

At home alone without Max I worried through another heat wave, another week cloistered in the relative coolness of my closet-sized home office, inside again, the smoke hugging the roof of Frank's house across the street. The wind grew to take up the space Max had left, accompanying me as I deadheaded flowers and fixed dirtbag meals and listened to audiobooks ramble on as though they were human company. With Max gone I had been having frequent nightmares—I blamed it on the wind—and there were many nights when the noise of air cascading through the trees found its way into my head. My dreams picked up the cue, making disaster movies: Floods. Tsunami. Boat wrecks. Sharks.

All my other dreams were of fire.

In one, a small fire in a tree by our house. The VFD boys get it but it's windy today and it spreads, ethereal. A bit later I'm out in the trees, on the path. Up at the reservoir the water is cool and I wade; minuscule azola ferns kiss my ankles. A military helicopter hovers, its huge basket lowers toward the water, I'm almost scooped up.

In one, Max and I are in a city walking on a median strip in the middle of a boulevard. It's hot. Traffic. There's a car empty, a woman in the car behind it is angry, she guns, revs, drives into it but there is no one in the other car to deploy the brakes. The two cars mate and tumble down the avenue, an action sequence. Somehow a plane has landed too, we are in it, it's like a cargo plane with a big open back part, the two cars approach, and everything explodes.

In one, fire besets me on multiple sides, maybe I'm in Santa Cruz in the mountains, maybe I'm at home, I'm in a car racing a smoke-black sky. I'm going ninety on a road, maybe it's Occidental Road, I aim for a large, flat meadow clear of trees, but in every direction are the flames; I'm already inside them.

In one, it's like it really is for so many people: I'm running from my home, wall of forest, wall of heat, the siren rings too late.

In one, the fire is beneath my skin, it erupts as boils at my pulse points, it's everywhere, except my belly, which is cold.

The much-publicized statistic that 95 percent of California wildfires were caused by humans came with a big asterisk. In California a surprisingly large amount of destructive wildfires were caused by a utility corporation, Pacific Gas & Electric (PG&E), which provided electricity to all of Northern California and much of Central California. (Southern California had a separate utility company, Southern California Edison, with a similar track record.) In 2019 an investigation found that over the span of six years—while the company was already on probation for negligence in a 2010 natural gas explosion that leveled an entire suburban neighborhood and killed six people—PG&E equipment started sixteen hundred fires, including the 2018 Camp Fire in Paradise, for which PG&E was later convicted of eighty-four counts of manslaughter. The company's oft-neglected equipment had a tendency, especially in high winds, to break, spark, and ignite any surrounding fuel—which was usually plentiful, as the vegetation around power lines tended to be poorly managed. PG&E was a public utility corporation that was privately held, which meant the utility was run for profit and managed by shareholders. It was also a monopoly. In the United States, where corporations were legally considered people, PG&E's outsized involvement in wildfire starts didn't technically change the math of that fire statistic. But it changed something inside me when I looked at my electric bill each month. The cost of electricity seemed to rise whenever PG&E lost another lawsuit or was convicted of more crimes. Every time I clicked Send in my bank's bill pay interface, I imagined my tens of dollars weaving through the virtual air like embers in a windstorm, igniting more fires. The cycle was endless. We paid the company, the company paid its shareholders instead of upgrading infrastructure and safety measures, the arm of the law weakly reached out to tap the corporation on the wrist via fines or class action lawsuits, and the corporation went back to the public—the same people whose homes and lives were being destroyed by fires, and who had no choice in their electricity provider—for more money. As it was, it had long been. On an old fire map of Sonoma County I'd found, created before the current naming convention, several of the wildfires from the 1960s were simply called PG&E Fire # ___. Fill in the blank.

It appeared the only way they could prevent their equipment from starting fires was to turn it off. In 2018, PG&E had begun a policy of cutting the power in certain areas whenever fire danger was high. The lines that caused the Camp Fire disaster weren't part of a shutoff, and the fire was sparked when a screw that hadn't been replaced since the 1920s failed and a live wire came unhinged by wind. In the following year, 2019, during a week of severe fire risk, PG&E and utility providers in Southern California instituted widespread outages, cutting power to as many as three million people. Max and I, after evacuating from the Kincade Fire along with ninety thousand of our neighbors, returned home a week later to the disheartening task of throwing away the entire contents of our refrigerator. The company offered customers $35 for food replacement costs—if we filed a claim with our insurance company, which for me would take more time and effort than was justifiable given the recompense. We stopped keeping meat in our freezer after that.

So far in 2020, the elements had shouldered the blame for the worst fires, and in the meantime PG&E had rebranded their forced outages as PSPS—Public Safety Power Shutoffs, an acronym that was impossible to pronounce or pluralize and that nobody I knew actually used. PG&E said that they, in partnership with state authorities, used a complex series of metrics to determine when and where the shutoffs would come, which, I once said to Max, hopefully included a data point for their busted-ass equipment. Generally speaking, the shutoffs were triggered by red flag warnings, which were issued when fire-weather conditions became severe. In an unincorporated area like the one I lived in, no electricity didn't merely mean no lights or TV. It meant no internet—the only provider was the cable company. And in the woods, no internet often meant no cellphones; service was spotty and a lot of people, myself included, relied on Wi-Fi assist for a signal. No power meant no emergency alerts. No siren, either. In the dark, people were without information, refrigeration, or hot water—or water at all, if you had a well.

Growing up in Santa Cruz we had something called earthquake weather, which wasn't a real scientific thing but people believed in it anyway. Earthquake weather, often arriving in warm fall months, was thought to be dry, windless, and carry with it a quality of uncanny quietude. There may not have been a distinct fire season anymore, but fire weather was a real thing—a combination of heat, wind, the humidity in the air, and the humidity of vegetation on the ground. Whenever the winds were predicted to rise and it wasn't raining, the weather service and local news issued fire-weather warnings. The warning system in Northern California was confusing to me even after years of experiencing it, despite the fact that I could tell when it was fire weather by the tilt of the wind and the shimmer of heat in the trees. Local news and state agencies circulated low-resolution maps of the warning zones. On the maps, pink, red, and orange blobs of color covered the relative risk areas. There were no road names and no way to zoom in closer to see whether one's town or house was included in the area of caution. Phrases like "the north coast mountain ranges" were vague; I lived amid hills that I would hardly call mountains, but they were connected to the coastal ranges, of which there were more than one. The only certainty when it came to fire weather was that people in Sonoma County would start worrying.

It was around this time that I began to notice fewer people saying things like, Is fire season over yet? Or, Can't wait for things to get back to normal next year. My neighbors were starting to slowly accept the nature of the ecology in which we lived. Fire season was no longer deployed as a unit of measurement indicative of the chaos of existence; now all anyone was talking about was living with fire.

This could mean different things to different people. It might include learning more about fire-adapted ecosystems and what they need to be healthy. For those who had already experienced fire, it might mean rebuilding with so-called fireproof materials. For many it would come to mean leaving, moving to a locale blessed with slightly less urgent evidence of climate change. Living with fire meant new hazards for workers, and new businesses: The commercials for cabinetry or HVAC services that I heard on the car radio were now using the word rebuild where previously they said remodel. (A similar change had occurred for a few months after the 2017 fires, but this time the lingo shift was permanent.) I began to see signs hammered into every telephone pole in West County, where it was already common for journeymen to advertise their day-labor, fence-building, and hauling services. One sign read simply FIRESCAPING, with a phone number. Others had been edited or amended: YARD WORK / HAULING / DEFENSIBLE SPACE. This was living with fire: living alongside it, in preparation for it, and in its aftermath, all at the same time.

In the last week of October, there was a fire-weather alert: red flag warning. Then came a fiercer warning: the wind was predicted to be worse than the worst times, worse than 2019, 2018, 2017. (Lately, lists of the worst times simply looked like a chronology of time.) Max was still electioneering in the desert. When PG&E announced PSPSes in my service area, I decided that spending days alone in the dark in high winds without cell service would be one stressor too much. It was time to go. Again.

I packed the car, bored with the ritual. My neural pathways were too tired to summon their fear synapses anymore. I was over it. I was angry. Energy corporations, governments, fossil fuel producers—everyone had known this crisis was coming. They had known how to avert it. They had done nothing. They were going to keep killing us, and nobody would stop them.

I pointed the car south toward the shaky corners of my memory, a time before disillusionment, and made for my hometown.

In Santa Cruz it was hot but not very windy. The CZU Lightning Complex was still burning, and would burn deep into winter, but it was contained. People in the Santa Cruz Mountains and on the coast were allowed to go home, if home was still there. Over the mountain pass heading west, I looked straight ahead. Distant ridgetops seemed to dangle from the sky, a looming grief. I knew it was burned badly up there. Up there, people didn't have houses anymore, and redwoods didn't have crowns. Santa Cruz had a lot of disasters, but this one felt bigger than the usual recover-rebuild-remember cycle that was by now familiar to anyone who lived anywhere the elements sometimes acted violently.

I didn't talk about it because it sounded so New Agey, but on this visit Santa Cruz felt to me as though the protection of the trees had been withdrawn. Without them the town was vulnerable. It was as though when the fire had moved through the redwoods, everything those trees had sheltered was exposed. Max had a theory that redwood trees were the closest thing contemporary humans had to a pantheistic group of gods—they'd seen it all, beyond our lifetimes' imaginations, and they watched humanity over eons as we floundered or soared. Without the presence of the gods there was no respite from the glare of the sun or the onslaught of the ocean's rise. On the way into town, encampments of people living in trailers lined the highway, the origins of their displacement unclear (pandemic, fires, housing crisis, mental health crisis, opioid crisis, take your pick). Police and ambulance sirens blared in the night. Half the shops downtown were boarded up, and in the other half the sad mid-Covid theater of commerce plodded into its next act. Before, there had been more oxygen here.

My parents had resumed their Covid safety protocols post-evacuation, as their proximal emergency was now past; I hadn't been inside their house in almost a year. For my self-evacuation, I managed to find a vacation rental in the garage of a retired woman who lived around the corner from them—one benefit of being from a tourist town, I supposed. It felt wrong to stand on my dad's concrete front steps, now overflowing with rows of pumpkins for the season, and not open the creaky front door. But the parents hosted me and my sister Caroline for a backyard dinner, at card tables spaced six feet apart. Orin was in good cheer, having recently been let back onto campus to tend the garden. Max chimed in from the desert, texting us pictures of dried-out blooms from the public rose garden in Reno. My dad identified each variety and gave his rundown of its strengths and weaknesses. He then brought out some good IPAs he'd been saving, an activity he and Max usually enjoyed together, and we took a bad group selfie to send back to our correspondent in the field. The evening felt a bit like a reenactment of our August evacuation, except we were more seasoned now, pros.

That night in the windowless bedroom of the vacation rental, I slept with my phone, unable to give it up although I knew the siren wouldn't call on me here. At home, the power was out. The wind was blowing. All one could do was wait. My pain was high-voltage; the current of its electric burden ran steady beneath my pulse.

In the morning there was no new news. I checked in with Max and we joked about which foods might be going bad in the fridge. Stephanie knocked on the garage door; she was driving up the coast to look for black swift nests, did I want to come? My stepmother was an artist whose work focused on California flora and avifauna; she had taught me more about the outdoors than anyone I'd ever known.

We took separate cars. North of town, the highway bordered the coastline. There were few houses; farmlands jutted out above the Pacific on clifftop plateaus. From the road I could see the Santa Cruz Mountains, now to my east, shadows above the coast. On each ridge the silhouettes of burned trees stood like vertebrae exposed.

Stephanie and I parked in a dirt turnout and took a short hike across train tracks and down a crumbly path worn into the mudstone hill. When our feet hit the sand, instead of making for the tide line like normal beachgoers might, we turned inland and headed for the backshore, the part of the beach that lay directly below the cliffs. There, below a curve of the rock that muted the noise of the waves, was a small human-made pond, irrigation storage for a farm up on the cliff. A pond was a pond, though: a family of ducks sat lined up on a log, plump and at rest. Steph and I stopped and turned in a complete circle, to admire the view. Clear blue ocean, clearer blue sky, lowish tide. Waves offered neat tubes to a handful of surfers. North of us the beach disappeared and grand cliffs made of stacked sandstone took the tide line position, curving along cove after cove, all the way to Canada.

Steph gestured at the rocks and ruddy cliffs, habitat for boundless species. It's hard to look at all this, she said, and believe that California is dying.

I couldn't tell if she was being ironic or not. Stephanie and I often joked about the hyperbolic language the press used when covering California's frequent natural disasters. We both knew that California wasn't dying; it was changing. Not always for the better. But it did feel like these days were the end of something.

Well, there's no place I'd rather be killed by, I said, and Stephanie laughed.

Seven miles up the road from here, the CZU Fire had come to these beaches. On its descent from Big Basin it forded the Pacific Coast Highway on a narrow curve of the road, moving west, and vaulted from blacktop to chamise. From there it bushwhacked through oceanside scrub and squeezed through cypress trees like a night wraith, haunting the low vegetation of the dunes before skimming across the water-filled blades of invasive ice plant that covered the cliffs, and there it leaned over the land as far as it could before the water denied it entry. I wished I could have seen it, flames meeting waves. The elements were working together as they always had, in a way hidden from human understanding. They were finding new ways to interlace, fire reinserting itself among them as though it had never been left behind.

Stephanie and I climbed up a steep path behind the irrigation pond, past aging pipes hoisting water up to the fields from a pump house where egrets often perched. On the clifftop we stood in bright, cold wind at the edge of a field of brussels sprouts. Using a small path we tacked around the field, which was private property, and eventually came to another inlet in the coastline. There, Stephanie pointed out the well-worn toeholds and indentations of a route in the rocks that led us back down to sea level, to another cove. The dark of a cave mouth was visible in the cliffside, and water churned into and out of it with the tides. This was where she'd seen the swifts, one foggy dawn. But from afar we saw no activity. You could go inside the cave, Steph told me, if you timed it right with the tides.

We headed back from the cove along the clifftop the way we'd come. Standing amid the crop rows along the highway, we stopped to admire the shoreline view.

Stephanie pointed out a rounded land feature a few coves north. It was an Ohlone midden, a sacred site. *Ohlone* was the general name used for the eight distinct language groups spoken by Indigenous people of this region at the time of colonization, and it was also a term used generally to describe the people themselves. The current-day Indigenous group in the region, the Amah Mutsun Tribal Band, was made up of descendants of people from the many tribes who were taken to the Santa Cruz and San Juan Bautista missions. This section of coast had once been home to Awaswas-speaking Indigenous communities including the Quiroste and Cotoni, people and languages now lost. The Amah Mutsun took responsibility for the care of this area. Here, and in forested locations inland such as San Vicente Redwoods State Park within the CZU burn area, the Amah Mutsun were partnering with California State Parks to reintroduce traditional resource management practices that included pruning, tending, and burning. A few miles northeast of where Stephanie and I walked, Amah Mutsun community members were part of an interdisciplinary research effort involving anthropologists, archaeologists, and scientists from UC Berkeley, UC Santa Cruz, and the San Francisco Estuary Institute. The research group was working to determine historical intervals of anthropogenic fire in this area, with the goal of modeling future managed-fire regimes. The Amah Mutsun Land Trust also partnered with a nearby educational farm that was owned by a white couple who had once been apprentices of my dad's. The partnership included access to the land for cultural use, training space for a Native youth land-stewardship program, and the establishment of a native plant garden at the farm. It was hardly "land back," the farm's owner later told me, but it was a start. Over the decades, Amah Mutsun members and their descendants had been priced out and dispersed from the northern Monterey Bay region; none of them lived in Santa Cruz anymore. They traveled here from other towns, often at great expense, to care for ancestral land. The night the CZU Fire began, the conservation crew had been staying in an old house on the farm while they worked in eucalyptus groves nearby. They evacuated along with everyone else on this part of the coast.

Before colonization, fire kept forests healthy; it also kept them in check. Much of this coastline had once been grasslands and coastal prairie, a now-rare ecosystem that was one of the most biodiverse in North America. The exclusion of fire from these ecosystems allowed the incursion of shrubs, Douglas firs, and ultimately redwoods, turning grasslands into forest, with detrimental effects to its available resources for people and animals. Whenever I drove the coast highway between Santa Cruz and San Francisco, I could see for myself evidence of these incursions in abundant clumps of pampas grass and teetering stands of eucalyptus trees—introduced species that were highly flammable.

Steph and I returned to sea level, passing the irrigation pond again. Excitedly, she pointed out four coots scooting across the surface of the water: oily ballerinas, their wings held aloft in tandem. Steph was always so exuberant when she was outdoors observing the natural world. Whether she was sighting a rare bird species for the first time or observing a common towhee, she was totally thrilled. I admired that in her: the capacity to keep joy that close to the surface. As I watched the coots walk on water, I allowed her ardor to infect me. I was filled with a feeling of admiration for the birds, the pond, this coast I'd taken for granted my whole life. What a privilege it was to live here. What a delight. My favorite trail on the Sonoma Coast was a cliff not unlike these; it was actively falling into the ocean. I wanted to be there when it sank, not in one clean chunk but in variegated ribbons of earth, rivers and crevices, a slow fall to the basalt floor of the Pacific. I wanted to be there when two different tones of California poppies grew on the disturbed soil, and I wanted to see ravens and hawks dip below the falling cliff's horizon in search of their next meal. If the forests were dying, temperatures were rising, and the lakes and rivers were running dry, I wanted to watch. What an honor it would be, to witness this marvelous land as it changed, until it no longer had use for me and sent me the way of the mammoths.

We were back on the main beach now, and Stephanie examined more closely the geologic layers of the cliffs while I gazed at the waves. At the tide line, a lone surfer prepared to enter the water. The swell was gentle, but offshore there were good, even-breaking waves. Pelicans hunted in the surf behind the break. The surfer waxed up his board with intent, not looking at the water. When he was finished, he flexed his leash, checked it again, and walked into the ocean like a messiah.

He joined a small coterie of surfers behind the break. They bobbed like seals in their wet suits, waiting between sets of waves. I had grown up watching my dad and brother surf, and I knew how difficult and dangerous it could be. I admired surfers' skills. And I loved the swagger of the sport, the sheer hubris of a human meeting an ocean inside of its boundless body, and making it into a game. I always wondered how surfers did it. How did they remain stable at their core while the churn of the earthsea roiled beneath them? Dip and bend, swerve and slide. I wanted to know how badly their knees hurt, how stinging the salt was in their eyes and ears. It seemed terrifying to me to place oneself at the mercy of the ocean, the largest and strongest force on the planet. I was curious about how it was possible to allow oneself to float willingly in chaos, and not let it overwhelm.

One of the surfers caught a sweet left break and briefly disappeared inside the tube before wiping out. He came up quickly in the shallows and headed right back out, facing the swash of waves. As the swell rolled in he punched the nose of his board through each wave, below the crest, allowing the water to push him backward for a moment while it passed, before resuming his paddle out. After each punch-through, he pushed his head into the wave, then surfaced like a playful otter. It was an action I'd observed a thousand times but I'd never before noticed how fun it looked, how blithesomely the surfer floundered on the surface of such power. It would have to be exhilarating, the liberty of an intimate relationship with an element: to eat and be eaten by water at the same time. The reason the water didn't frighten him was that he loved it.

A future memory. It is hot and sunny and springtime when I am allowed past a barricade, up a closed highway, and into Big Basin Redwoods State Park. The park, established in 1902 in the Santa Cruz Mountains, is beloved by generations of Californians. It has two entrances: one in the inland mountains and the other on the coast, about six miles away as the crow flies, or thirty miles on human roads. Between them is a bowl-like swath of eighteen thousand acres of redwood and mixed-conifer forest, a quarter of which is estimated to be old-growth. This is Awaswas land.

The park is still off-limits to the public but a senior scientist for California State Parks, Joanne Kerbavaz, has agreed to show me around. Joanne is a friendly but pragmatic white woman with the sun-wizened skin of a person who works outdoors. Along for the tour are a botanist with the California Native Plant Society, whose name I don't catch, and a retired ranger named Mimi, who lived in the park when she worked here and still volunteers as a docent. It has been two rainy seasons since the CZU Lightning Complex Fire, neither of which was very rainy. After the CZU conflagration, fire lingered in some tree trunks and stumps for as long as a year, sometimes rekindling and creating new fires that had to be extinguished. Fire can also persist underground, fueled by the oxygen and organic matter, such as roots, in the soil. Where I am, in an old-growth stand of redwoods near the former visitor center, the ground is cold. It is very quiet.

California is supposed to burn, but not like this. Even for a fire-adapted landscape, the 2020 fire was too much. Like many of the increasingly destructive wildfires the western United States has experienced since 2010, it burned too hot and too fast. Fire came from three directions, so wind-driven that it climbed into the crowns of trees that were hundreds of feet high. Ninety-seven percent of the park was within the burn area, and most of the human-made infrastructure and buildings were destroyed, campgrounds and all. (*Park-itecture,* the rangers call such non-flora features.) Before the CZU, it had been at least a hundred years since Big Basin had burned on a large landscape scale. The oldest known redwoods in the park are about three hundred feet tall, and they are thought to be one to two thousand years old. Because history and the oral inheritance of Indigenous land knowledge have been broken by genocide, and because the current warming of the planet is unprecedented in human experience, nobody now living really knows what will happen to old-growth redwood forests after fires like these, let alone whether what happens might be valued as good or bad. When the park reopens, the people who love this forest will say it is unrecognizable, forever changed. Heartbreaking. And what has been broken will remain broken. Meanwhile, the forest knows what to do.

It's early morning. The only animal sounds I hear are the constant hammering of pileated woodpeckers, who thrive in postfire forests. Around me stand the giants. They look different now. They smell different. The forest is making a new version of itself. In this iteration the redwoods are mostly still alive. But they have no branches or needles. Their godskin is shorn smooth; their charcoal bark makes marks on deer noses and human hands. I look up, and I realize that the giants' structure is now clearly visible from below—no branches, remember—and so the trees look taller and thinner than I remember them being. They are severe yet somehow also coltish, in the way that very old people are when they experience the regression of late life. Soon the sound of chain saws from a maintenance crew joins the sound of woodpeckers and they sing together of how different, how hurt a place can be.

We walk the old-growth loop. Burned fences and signs, warm dry dirt. The Mother Tree—so named when the park was first opened, back in the days when to love a forest meant to freeze it in time—is undercut by a deep vee of negative space at its base. Still alive, Joanne says. She shows us two tall tan oaks that form a neat arch over a path. These trees' leaves are extant but dead, heat-shocked by the fire. At first, we thought the trees were dead, Joanne explains. It took about eighteen months, but they resprouted.

The cat-eye triangle that is burned into the base of the Mother Tree predated the 2020 fire, but it is larger and deeper now. I stand inside it and crane my head up toward the solid core of the tree, which begins at least another twenty feet above my head. When the sunlight hits the microscopic particles of charcoal that still hover in the air inside this cavernous trunk, it looks like a rain made of light, like starshine ether, like air itself.

With the loss of the canopy it will be hot and sunny beneath the redwoods now, for a very long time. The forest floor is full of oily green bushes of sun-loving ceanothus, newly grown. Nitrogen-fixers, doing their job, the botanist says. It's remarkable to see them under redwoods, says Joanne. The shrubs would never have thrived here when the canopy still shaded the floor. Now they devour the newly available light. Huckleberry, a longtime resident of the redwood understory with delicious, tart berries, grows in unruly bunches alongside its new neighbor.

We hike along a creek bordered with fallen trees, known as deadfall, and burned vegetation, clearing a path as we walk. In a conifer forest, deadfall and standing dead trees, called snags, serve as valuable habitat for other plants and critters. The botanist photographs a fern that none of us recognizes. I dangle my hand in the water; it's cold. I stop to admire a few redwood seedlings in the dirt, each barely a handwidth high. Yeah, Joanne says, seedlings are cool. But sprouts—the shoots that grow from existing plant structures, as opposed to from seeds—are stronger. Sprouts have a much better chance of survival.

We drive in trucks to another part of the park and stand on the roadside looking down at the burned basin: a sea of mighty toothpicks. In less-accessible parts of the forest, where there has been less human recovery work, the fire's destructive power is more palpable. Redwood trees, in addition to resprouting from their roots, are capable of putting out new growth from every available area. If they still have branches after a fire, they will sprout from those branches, although most of the burned redwoods I see don't have branches. They do, however, have trunks, and a chaotic fuzz of new needles runs up and down them. They look like bottle brushes, everyone agrees. The plant scientist demurs: Actually, I think it looks like they're wearing little sweaters, she says shyly.

Deeper into the park, we stop at a large meadow. *Calochortus albus*—fairy lantern flowers—droop in soft pinks and whites; native irises display fuzzy yellow and purple tiger stripes at the nape of their petals. The small blue flecks I see in the meadow are blue dicks, a common wildflower. I stroll for a bit with Mimi, the ex-ranger who sometimes volunteers as a docent. When visitors return, she says to me, they're going to ask, When will it be back to normal? That depends on what your definition of normal is, she says.

I say, I think what people are asking is, when will it grow back to the way it was before?

She waves her hand at the burned redwoods surrounding the meadow and replies, What way? The way you like it? The aesthetic image you have in your head of what a redwood forest *should* look like? That's not what the forest is right now, she says.

She seems already tired of navigating the grief of the general public, and the park hasn't even reopened.

What are they looking for, a number of years? she asks. We don't know.

Before the fire I had been to this park many dozens of times throughout my life. But now that I'm here, I can no longer envision what it was like before it burned. There is no way to compare the two forests. She's right: they are different selves.

After the initial news cycle of the 2020 fires faded, the press releases and articles updating the public on the regrowth of Big Basin have been largely hopeful. The messaging focuses heavily on postfire renewal, the strength and adaptability of the redwood forest, and the importance of reopening the park to visitors. The public is urged not to despair, and instead to make donations to the rebuilding effort. (Because of funding cuts, many state parks in California are assisted financially by partnerships with nonprofits.) The narrative of renewal appears angled to reassure a public that feels the loss of this place profoundly. But fire is nuanced. It can be a renewer, a tool, or a killer. An apocalypse or an ancestor.

In general, the progress of any postfire landscape is uncertain. Postfire ecology is as complicated as other ecologies. There are no rules and no absolutes that apply to burn areas. Like plants themselves, the science is constantly evolving. Severe events can change ecosystems. When landscapes burn, habitats can move; patterns of sunlight, water, nutrients, and predation might shift, proving beneficial for some species and not for others. Not all species make it through. Fire-adapted species often display new vigor, but others never come back. Some things in this park have died entirely: redwoods, yes, although their adaptations make their overall survival rate high. Virtually every Douglas fir has been killed by the fire. No one has seen any chipmunks yet; they used to stalk the picnic tables like mosquitoes. After the fire, in 2021, a marbled murrelet nest was spotted in one of the few remaining Douglas firs in 2021, inspiring a flurry of publicity. MAMUs, as they are affectionately called by bird nerds, are rare, enigmatic, and threatened. They live at sea and nest in the canopy of old-growth coastal forests; Big Basin is the southernmost reach of their habitat. The nest and its successful fledgling were a positive occurrence, but the return of MAMUs to this forest is uncertain: one nest does not a repopulated species make. The birder who initially discovered the postfire nest, Alex Rinkert, explained to me that MAMUs prefer thick cover; this fire-damaged tree had hardly any foliage left, and it was uncomfortably close to an area that would be crowded with visitors once the park reopened. It was unclear whether the marbled murrelet's singular appearance signaled a postfire miracle—resilience! life finds a way!—or an existence that is less metaphorically convenient. It's complicated.

Our little tour party caravans up bumpy, half-melted roads to an inland ridge, where it's much hotter. The view is splendid; the ocean appears to sparkle in the distance. No redwoods grow up here. At this elevation the park is chaparral and sparse pine forest. Sooty patches of soil stretch between large sand-colored rocks. I ask Joanne if soil can die. She doesn't know. She hasn't seen any dead zones in the park. Low-intensity fires can actually add nutrients to some soils, but high-intensity wildfires are capable of partially sterilizing soil by killing the microorganisms that contribute to its organic makeup. I see other dead things. New manzanita sprouts grow next to the twisted remains of their parent trees. Manzanita hybridizes freely, creating new varieties through reproduction; I wonder if the fire here sparked new types. Up here, more opportunists abound: thistle in incessant purple spikes, ubiquitous broom. A bright yellow bush poppy reflects sunshine back to the sky. The landscape is strangled in white morning glory.

On the empty road, pine cones roll underfoot. Nearby are the burned remnants of knobcone pine trees. These trees are the *real* fire story, Joanne says: knobcone pines are obligate seeders, and their cones won't open without fire. What's an obligate seeder, I whisper to the plant scientist, and she makes a scattering motion with her hand, as though she's flinging seeds on the ground. There should be approximately twenty-five years between fires in knobcone pine forests, to allow trees to mature enough to produce cones and seeds. But even with fire-necessary species like these, it doesn't always go so well. When it burns this hot, Joanne says, that quarter-century seed cycle could be wiped out in one night.

I'm impressed by the postfire life of Big Basin, but I can't deny that it's devastating to be here. Before now I've been in burned parts of forests and towns, but I have never been entirely inside a wildland fire scar, and never in an environment I know well. Alongside my grief, I feel something else. My prior assumptions—that a burned place is either a moonscape ruin or a rebounding font of organic life, but nothing in between—are becoming quickly confused.

In the book *Islands of Abandonment,* journalist Cal Flyn visited sites that had been abandoned or injured in some way by humanity—toxic waste dumps, islands left unpopulated by environmental disaster, radiation exclusion zones that were now overrun by wildlife. In some ways, Flyn writes, the places that are recovering from harm by humans—places thought to be dead—are more "authentically alive" than other, traditionally touristic natural places. She notes that oftentimes, what people assume is a truth about a place is actually a judgment based on their specific cultural viewpoint. In order for people to live in the damaged environments that climate change is making, Flyn advocates the cultivation of "a new way of seeing: a new way of looking at the land." When I stand on the ridge in Big Basin and gaze out at the vast black treescape adorned with fuzzy green sweaters, I am reminded of a shelf of books I've collected at home. They are books about fire, and their covers are red. Red and orange. Flames and more flames. Bright and hot. But this burned place shows me that the color of fire is not red. It is black: dirt pulverized by heat. It is green: new laurel sprouts and carpets of ferns. The color of fire is blue and white: sky and clouds that can be seen easily from beneath a thousand-year-old tree with no branches. I might recognize it as beautiful if it didn't all feel so askew.

After an entire day spent standing in the ruins of redwoods, I take the long way back through the mountains to the coast. I drive slowly, I don't see many other cars. Gillian Welch sings about California on the stereo. The road takes me through neighborhoods that look a lot like the one I live in back in Sonoma County: small wooden houses and quirky little towns hidden amid forested hills. The CZU Fire came here, too, and even eighteen months later I recognize its imprint: mailboxes where there are no houses. A telltale driveway with a blank patch of earth as its destination. Trailers next to piles of bricks. I drive through the unincorporated areas of Boulder Creek and then Bonny Doon, where I was born, and I try to remember what it used to look like. There are small, almost secret utopias in these hills: former hippie compounds at Alba Road, Swanton Road, and other formative locales that I claim as part of my origin story, though I haven't been up this way in years. I roll down the window and the car is filled with the earthen smell of the forest, so oddly absent when I was in the park.

A MEMORY. My first. I am a toddler sitting in dirt, knees bare, head full of the smell of redwoods and eucalyptus. I wipe my nose and red dust comes out: tree sneezes.

I remember sitting in the wayback of a station wagon on a road exactly like this. I remember the downslant of sun through branches, the branches big as trees, trees bigger than the sun, the road twistier than my stomach. The rhythm of the shade on my skin keeps pace with the creek's whisper.

I can stand inside trees.

I remember a little red schoolhouse, a commune where midwives lived. Wooden houses, some finished, some still being born. Crooked angles. The tang of cast iron in the kitchen. A plastic, aboveground pool filled with dirty warm water, gangs of kids at play in the dirt. Leaves of three, let it be. Plantain weed is for leg scrapes, chew it up and press it on the wound.

I remember being carried to the car and how cold the night feels but how soft.

The howl of feral rain.

MY MEMORIES of these trees and mountains, before the fire, are inextricable from my childhood, my identity. With absence, the physical sensuality of any locale fades; over time, the experiences and places that form a person come to dwell mostly in memory. As decades pass, a person's primary relationship with the places of their origin becomes more about the idea of the places than their physical reality. With fire my experience of this phenomenon becomes more pronounced. A burned place confounds memory. Fire makes the familiar newly strange. In the mountains, after fire, I am unsure of how to deal with the knowledge that these parts of my past no longer exist within a physical space. My memory of the place has become the place.

When will it be back to the way it was, I repeat out my car window to the cruel blue sky, a refrain of grief. Not within my lifetime. Possibly not ever. And who am I to say how a forest should be? I will learn to see beauty in the black and the green.

· 10 ·

Veils

While I was self-evacuated in Santa Cruz, the October winds didn't deliver the promised destruction in Sonoma County. The red flag warning was canceled, the power came back on, I went home. There, I surveyed the garden, after wind. The meadow and the beds were brown, not because their soil was brown but because they'd been covered in blown redwood needles and oak leaves. Sticks the size of my arm protruded from bushes. The flower beds looked like a forest in themselves, small whips of fallen wood amid the perennials. There were meter-long pieces of redwood branches in the hedgerow, and a few lavender bushes had taken hits. One of the fallen branches in the yard was as big as a small tree. Another widow-maker had apparently plunged straight down, vertically: near the fallen branch was a surprisingly deep hole in the dry, hard ground.

I picked up branches. I picked up sticks. I wrangled rakefuls of oak leaves as big as my head. I swept duff from the back deck with the mangled old broom we kept in the potting shed. I pushed roses and jasmine and fuchsia back to standing positions. It looked as though the yard had been through something, but it was hard to say what. The wind had had free rein all weekend. Trivial and not-so-trivial destructions. Relatively speaking, we had come through the fire weather unscathed. I still felt sad.

In fire places, a common small-talk ritual among residents was for a person to ask, Were you affected by the fires? This was a polite way of asking if your house burned down. The customary reply was, Oh, we evacuated but we were fine. What I always wanted to say instead was, Were you *not* affected by the fires? Did you live on this planet, on this land that was begging to be burned and at the same time burning too hot? Did you breathe the air around you? Did you ever feel love for a plant, for a tree? I understood that people mattered but it was the trees that I grieved. I had to believe that if you extinguished in violent heat the life force of thousands—millions?—of living, breathing flora and the fauna that lived inside them, it mattered. There was a collective sob heard and felt within. I could understand that a forest would grow again after a fire and still mourn its inexpressible losses.

After I was done picking up sticks, I made a pile of them in the middle of the yard. The best way to dispose of so much organic matter was to burn it, but the wind still blew. There would be no backyard fires for a long time. The pile resembled an osprey's nest.

I clasped my hands in the air and stretched my back, looking up at the canopy. The wind, calmer but persistent, cut through the trees and into me. I allowed myself a moment to feel my body's sensations. My skin was dry, my mouth and ears felt coarse inside; my *brain* felt dry. Since the wind had arrived, every morning and night I rubbed moisturizers and salves into my limbs and belly, but the little corners of me stayed brittle. My ligaments were frazzled. My pain today felt like negative space.

I was by now accustomed to the grief I felt for the loss of my body parts. It was not the grief of a specific loss—losing the ability to reproduce or the other functions those organs performed. Instead my grief presented as a faraway sense of unwholeness, the confusion of a biological being missing some central component. In the garden I had seen birds whose nests had been destroyed hopping on the ground, murmuring irregular songs—this was called displacement behavior. At times it felt as though my body was performing its own shuffle of displacement, trying to find new ways to feel at home. The grief I felt for the landscapes of my life, although it had been circling my awareness for decades, was not yet familiar.

Since I was already out in the garden I decided to displace a few weeds while I was at it. It was killer time. I bent at the waist and groped wildly at clover and dandelion in the beds. (Dandelion, a culinary delicacy and powerfully resilient plant, was allowed to grow freely in the meadow, but in the flower beds I was still old-school and weeded it out.) The farther I bent over, the more my pain drew me in, until its space became a whirlpool became a black hole, a tear in the universe. I took a seat in the dirt. From there I could twist and reach far enough to drag a weeding tool between plantings to catch the tops of weeds; those large carrot-like ones were impossible to root out. The stance hurt my back, but it was a refreshing break from the harrowed feeling of my pelvic muscles. I put down the tool and grabbed on to a vine of sweet pea protruding from beneath a rock wall that divided the beds. Sweet peas appeared delicate but they were also a killer in this garden, strangling other plants. They grew wild all over West County; we let them bloom in spring, then weeded them in the fall. I yanked on the vine and it rippled like an eel as I pulled, unearthing a line of root that stretched beyond the bed I was sitting in, beneath a rosebush, up a small rock wall, and into the upper tier of the garden. I was forced to stand to keep tugging, and when I did, everything inside me felt similarly raked across the surface of the soil.

A friend once told me about the work of psychologist Joanna Macy, who was one of the early practitioners of what came to be called ecopsychology. Macy, a Berkeley-based psychologist who was now in her nineties, spent much of her career working with grief. In a 1995 essay she described a phenomenon that she was seeing more frequently among patients, a feeling that she termed "pain for the world." Such pain—call it solastalgia, grief, that amorphous lack, the empathetic sadness that one feels when apprehending the scale of climate change—was a good and necessary pain, Macy argued. Pain was the correct reaction to what was happening. Pain for the world was real, valid, and healthy. In order to move forward as people in the face of planet-level change, pain was necessary. The fact that a person could feel pain for nonhuman systems showed that people were in fact part of those systems. Humans were connected—physically, emotionally, spiritually—to the planet; it was natural that we would feel its pain. It was essential, said Macy, to sit with that pain. "Be its gardener," she wrote.

I gave up on discovering the source of the sweet pea vine, and instead I walked through the wind-rumpled beds looking for marigolds. It was almost Day of the Dead, the Mexican holiday that followed Halloween. At this time of year there weren't many flowers left blooming in the garden; the sun had moved lower behind the tree line earlier each day, in a way that felt permanent. Marigolds were late bloomers and toughies. We interplanted them everywhere because they were guaranteed blooms in the fall, and Max had heard they were natural pest repellents. In Mexico and much of the Latin American diaspora, marigolds were the joyous markers of Día de los Muertos.

I brought a handful of marigolds inside and began to put together an altar on the mantel above the fireplace. I'd saved my burned bay leaves from August; I made a rainbow of them, from tan to char. I took a couple of the marigold blooms and snapped their stems off, adding their sunny orbs next to the array of burned bay leaves: a sun and its dark moons; the dead and the recently living. Photos of deceased loved ones came next—Rhys's dad, my grandparents, Max's uncle who died by suicide when Max was thirteen. The altar was a ritual I'd adopted during the fifteen years I lived in San Francisco's Mission District, then a gentrifying Latinx neighborhood. My mother had always kept a small altar to the god Vishnu, and I had been making my own, more punk-rock altars to friends, artists, and musicians for much of my adult life, but not until I lived in the Mission did I realize how heavily my practice had been influenced by—or, put another way, appropriated from—Latin American cultures in California. In the Mission, on every Day of the Dead the neighborhood park filled with altars, each of them works of art, and at night a street parade, jubilant and spooky, meandered through the streets. I valued the occasion as a dedicated moment to welcome the dead back into my life, to remember them, and to find joy in the revolutions of the years since they departed. The holiday was a rare opportunity to sit with the knowledge that people died; people I loved had died; I, too, would die—obvious facts, but ones of which it felt increasingly difficult to find public cultural acknowledgment. I added an apple to my mantel, for the harvest and sweetness. For beauty, a rose in a bud vase.

Next I rifled through my tarot deck. During inside days, while Max had been getting into the Torah, I had been getting into tarot. The tarot was nominally a card game that over many centuries and across many different points of origin had emerged as a sort of soothsaying practice. Along with astrology, it had recently been making a comeback in popular culture. The game was simple, although there were infinite variations and spreads. You ask the cards a question; the card you draw is their reply. The tarot was an exercise in close reading. The deck's iconography dealt in characters and scenarios. I thought of it less as a divination or spiritual practice than as a storytelling tool, although admittedly the game's appeal to me was heightened by its being a woo ritual (as Rhys would say) that was in no way associated with my parents. The tarot was all me. I used it as a creative exercise, mostly, a way to direct my day's work in a more focused manner. I pulled one card a day. Sometimes I drew cards for characters and sometimes for projects. Most mornings I asked the deck simply, What do I need to know today? I never asked to foretell the future. It didn't work like that.

Today I cheated, and I looked through the deck until I found the card I wanted: Death. A skeleton holding marigolds. In tarot practice, death meant change.

Now, the elements: An acorn from the oak tree in the yard, for earth. Air was represented by an orange feather from a northern flicker Max and I had found dead on the path. I filled a small vessel with water and arranged it between them. It needed fire. I grabbed two candlesticks from the kitchen table and framed my tender objects with them. Spark, flame; heat, light. Wildfire was my abiding specter, but it was not the only thing fire made.

I stared at the candle, then closed my eyes to it, and silhouettes of flames shaded the inside of my eyelids. The dead trees came back to me, miles and miles of incinerated lives. Maybe these orbs of golden musk—flowers and flames—would summon the spirits of the burned. Maybe they could help bring balance back to the seasons. I extended my hand and held an index finger above the candle for a few seconds. The flame warmed my skin rapidly, tantalizingly, while inside me it felt as though a sheet of aluminum wrap had been stretched across the hammock of my pelvic floor and was slowly being warmed until it roasted me from within. I wondered what would happen if every time I felt that ache, I thought of it as pain for the land, not myself.

Joanna Macy, the grief psychologist, said that when people felt the pain of the world they tended to feel hopeless, like everything was falling apart. The Italian writer Elena Ferrante, one of my favorite novelists, had made the interior experience of falling apart a recurring theme in her work, and as I withdrew my red and tender finger I recalled the word she used for it, *frantumaglia,* a fragmentation, an interior jumbling of a person's perception of how her world is structured and where she stands in relation to it. In Macy's clinical approach, which she called "despair work," the falling apart was the important part. The act of fragmentation that occurred when people felt pain for the world allowed us to develop a new relationship to and understanding of our power. In despair work, power wasn't defined as something that a person wielded over another person or nonperson. Power, Macy wrote, was in practice "the ability to effect change." Power *to*. Power *with*. Like a good relationship, power was a shared exchange of care and energy that bettered both parties. That type of power required vulnerability; it required a sense of joy in the falling apart. As in a relationship, it required love. Grief was an expression of love, after all, and you had to let yourself feel them both, or you didn't feel them both fully.

It was a season of death in a year of death in a culture bent on ignoring death, within a national body politic that was consumed with the gobbling up of itself by a death cult called oil, capital, empire. The growing terrestrial darkness of late autumn was a time to honor and speak with the dead, to acknowledge the spiritual elements beyond human grasp. It was a time to dress in costumes befitting the myths of one's particular culture and venture into darkness unafraid. In farms and gardens the harvest was over and it was time to turn under summer's spent crops. Fields were left to dry out or were plowed into dirt. Piles of dead matter accrued between flower beds. The cycle continued.

For me, every moment was consumed with waiting. Max and I texted and talked most days, but I was lonely. I was waiting for Max to return so I could sleep again. I felt like waiting was all I did: waiting for the next fire to spark, for rain to arrive, for my pain to break down the barriers of the corporeal and push me deeper into the wilds of myself.

Something about the season led to the sense that the reality I was experiencing was not the only layer of existence. I had often heard the term *veil* used to describe a feeling of thinness that was common at this time of year. The veil was like a gauze curtain, a porous border between lived experience and experiences beyond life. Ten years earlier when my grandmother had been dying of Alzheimer's, my mom sent me videos from her bedside. My grandmother was not a particularly spiritual woman; she was a sophisticate and an art curator; she loved nice things. In the video she was in bed in her room at the residential care facility. She wasn't looking at the camera or at my mom. Instead her gaze was fixed on a point both near and far, as though she were admiring a landscape that only she could see. She was talking in a manner that some might call speaking in tongues: her voice trembled and rose in a babbling singsong that sounded not random or diseased but like language, only it was a language I couldn't understand. She's visiting the other side, my mom said, she's been going back and forth. The veil—that delicate boundary between life and death—was growing thinner. She died soon after.

I had found that people I knew who had experienced health crises tended to be more attuned to the subtext of mortality coursing through the living world. After being under general anesthesia for my first surgery, I said to a nurse, still high, that I had been to the other side. She laughed but didn't disagree. In the experience of having my heart and brain functions slowed to the point of nonresponse, I had temporarily slipped through the veil, or at least been veil-adjacent. With each procedure, every time my consciousness was offline, I came back from the other side feeling a strong sense of disorientation that often took weeks to go away. Like a bad dream that lingered in body and mind, though my recollection was nebulous I knew it had happened.

In the western woods, as the light fell lower behind the trees, the veil became thinner. Other boundaries wavered too. There were rumors of alarming developments in the state of on-the-ground politics. Max reported back from Nevada that people were being mean to canvassers in ways he'd never seen. He'd been yelled at during house visits by residents who rifled off insults and right-wing talking points taken verbatim from TV programs. One volunteer had to be sent home when he started saying violent things about women and declared himself redpilled. Out in the field, a white man in a suburb had pulled a gun on a volunteer, a Black woman, when she approached his house to discuss the election. There was talk among other friends who were canvassing in swing states: no matter who won, we should be prepared for some sort of mass violence on Election Day. And there were rumors that fascists might try to stage an insurrection, possibly occupying government buildings en masse. Whatever was going to happen on Election Day, it was obvious that the United States was marching toward discord. At the same time, politics, culture, and economic structures were barreling onward, refusing to acknowledge, let alone mourn, the dead of Covid. This denial of grief had dovetailed with the ongoing denial of climate collapse, that suicidal capacity of Western humanism that seemed bent on doubling down, betting everything on exploitative economies rather than admit it wasn't working. Health policy had become politics, politics had become religion, and nobody believed anything except what they already believed.

Meanwhile, Northern California was slipping toward a new season. The Glass Fire was reaching full containment, after consuming fifteen hundred buildings and almost seventy thousand acres in Napa and Sonoma Counties. Eight percent of the wine grapes in the region had been left unharvested due to smoke taint. Seventy thousand evacuees in the county had gone home. The air was still dry and the fire danger still high, but days began to feel crisp. My anxiety calmed with the weather. I harvested buds from the marijuana plants Max had been cultivating and was surprised by the subtle beauty of those flowers. I tacked lengths of twine across the ceiling of the spare bedroom/office and hung the stems like flowers to dry, upside down. For the first time in a long time, I drove around doing errands with the windows down. I bought bread and flowers from local spots and the car filled with their smells and Michael Stipe's reedy voice came on the radio and I felt like, Okay, I love living here. When I got home I found I could write, I could spend time offline, I could walk on the path without the pain edging in. That night, I found I could look out the window and imagine rain over my garden. I could sleep. On days like this it still felt like I might get away with this existence, in this place, as it was. The day before Halloween, for the first time in months it was cool enough in the evening to close the windows, and as I did so I felt the glass had become a barrier to noise, smoke, and the fearsomeness of the era in which I lived. I didn't do my nightly prepper routine; instead, I swaddled my head in a thin scarf and lay on the couch and allowed myself to sleep.

Kelly's text woke me.

Fire, it read. Here. On the path.

The windows had been closed, and I had been deep inside the soporific consciousness of a person who hadn't slept well in months, finally at rest. So I didn't hear the siren. I didn't notice my phone buzz with the alert from the dispatch app. I didn't see the social media posts from the emergency info accounts. I did not smell smoke. But the message from Kelly vibrated the couch, and as I shook dreams from my subconscious I heard an unsettling noise outside: helicopter blades like thousands of arrows slapped through the air above the trees.

It was almost midnight on October 30, the cusp of Samhain, All Hallows' Eve, and Day of the Dead: time for Persephone to return to the underworld. I got up.

I opened the front door and looked wildly out into the dark. I smelled it then. A quick, deep, acrid smell, with a thinner presence than the smoke I was accustomed to from afar firestorms. Our new fire was dirty and urgent. It was very, very close.

I had no way of knowing if the fire was small or large, if it was spreading or not, or where exactly it was. Kelly texted that she could see it, she thought, behind her house; from her bedroom window she could at least see several fire engines and men pickaxing their way through the underbrush up the hill: the siren's effect. Should we go, Kelly texted. There was no communication from any authority. That wasn't unusual; the county would only contact people in the case of an evacuation-level event. But emergency protocol never mentioned anything about the people who lived next door to the event. Who told them when it was time to leave? I looked around at my neighbors' windows. Lights were on in houses, which meant people were awake, but no one appeared to be taking action. Then I heard what sounded like heavy machinery, the backup beep of large vehicles, coming from high on the path. I threw a hoodie in the car, all my careful preparations instantly irrelevant, and I texted Kelly to meet me in town. This was it.

Night. Quiet. No copters in town. No people either. The blink of the sign for the arts center patterned slowly, brightly. Kelly was already parked across the street from the empty firehouse, her vintage red ride unmissable. Nootka paced nervously, leashed. He pawed me when I approached and I was grateful for the affection.

We waited.

Are you cold? asked Kelly, and I found that I was. She brought out two blankets from her trunk. We sat on the sidewalk, leaning against an embankment. She pulled Nootka in and held him on her lap, like a child. Her own child was at her ex's house that night; Kelly and I were both alone at home. I nuzzled into my blanket. The sidewalk was solid below me. My pain was a ghost, never at rest. Time fragmented and converged. Kelly and I took turns pulling up the social media account of an emergency management nerd who posted scanner-gleaned information about wildfires and who had become a reliable source of real-time fire info. I had learned this fire year that there was a whole world of people near and far with an obsession and technical resources. His most recent post, a few minutes earlier: *Moderate rate of spread; heavy timber.* Phone reception in town was always slow. We refreshed.

Kelly texted with other neighbors, who had no news either. I stared at the sky above the tree line, looking for smoke from the direction of our houses. Smell of pine, smell of car exhaust as a lone truck passed and turned east, not in any apparent hurry. Cold of concrete, fuzz of blanket. It was past midnight now.

Happy Halloween, I said to Kelly, and we laughed. Her favorite holiday.

Didn't it used to rain on Halloween? she asked. She grew up in the Bay Area too.

Yeah, totally, I said. As teenagers, my friends and I dressed for Halloween in skimpy costumes and went trick-or-treating in the rich neighborhoods, and nobody would bring a jacket. It would always start to rain. We froze our asses off every year, I said.

When I was younger my dad had taught me the difference between weather and climate. Climate was the way the atmosphere behaved over a long time period—decades, for example. Weather, on the other hand, described smaller, short-term conditions. Changes in one often resulted in changes in the other, but it wasn't always possible to prove direct causation. A favorite pastime of certain leftist pundits in recent times had been to argue about whether each extreme weather event that occurred could be attributed to climate change, or whether people were experiencing confirmation bias and simply noticing the weather more now, because everyone knew the climate had changed. I mean, they weren't climate deniers or anything, these guys would say, but correlation was not causation. We must be objective. Bros always want evidence, I thought when I scrolled past their arguments, but all human experience is ultimately anecdote. It seemed to me that anyone who paid the smallest bit of attention to the world around them could see clearly what was happening. It did used to rain on Halloween.

I stood and started stomping my feet to warm them. I remembered the crispy weather earlier that day, and I inhaled as though I could still sip it from the air. That taste of cool fall weather, a moment of normalcy within a climate lost to chaos: what a gift.

The basic weather-related equation that had overtaken my life—heat plus air plus fuel equals fire—was in practice much more complex than I first thought. In addition to wind and fuel loads, a key measure of wildfire risk was relative humidity, a ratio used to gauge different aspects of humidity in air versus plants. The humidity percentages I saw on weather reports measured the amount of moisture in the air, but it was much more complicated than that. Humidity changed as moisture moved back and forth between plants and air—an amazing fact, air and earth swapping spit like some elemental make-out session. If the humidity was low, air was more likely to absorb moisture from vegetation, and so the vegetation—the fuel—would be drier. If humidity was high, the air wouldn't absorb much moisture from the plants. That's where the relative part came in. I struggled to understand the intricacies of the math, but I got the takeaway: low relative humidity meant higher fire risk. In fuel-heavy parts of California, the vegetation moisture levels were measured at regular intervals; drier vegetation, earlier in the year, meant more and worse fires. In the conifer forest on the path up the hill behind the house in which I lived, the plants weren't thirsty tonight. Likewise, the wind happened to be weak.

The scanner guy posted an update: *forward progress stopped.* This meant the fire was not spreading, although it was still burning.

Kelly verified the scanner guy's report by looking at the neighborhood chat. We didn't know what else to do, so we decided to go home.

Our fire hadn't spread.

The next morning Kelly and I went to look at the burn scar. Our fire, which is what I named it, had been about halfway between her house and mine—a five-minute walk for each of us. We met on the path; Kelly brought Nootka and another neighbor, a thin white guy wearing a crisp T-shirt and jeans. The burn area was about two acres, beginning in a small fairy ring of redwoods next to the path and stretching up a steep incline. It looked like any other burned copse: red wood gone black, oak leaves shocked sepia. The redwoods' trunks were charred where they met the soil; about ten feet above my head they began a slow fade back into ruddy vitality. If I looked farther up, to the crown, it was green and blue up there like it should be. The fire had cleared the forest floor of duff. Across the path, where the fire hadn't reached, I saw a mirror of these acres' past: an understory of dead plant matter, five feet high. Impenetrable stands of invasives. Human trash. That peculiar feeling of desolation, of death, that I always had when I walked up here.

Did you see the creepy trailer? said Kelly. She was referring to the abandoned trailer farther up the path toward her place: DEMOLISH ME. Last night the fire department had bulldozed it when they came to fight the fire. It was now tilted over to one side, teetering on the crest of the hill downslope from the path, one axle on the trail and one axle in the bush.

As Kelly and the neighbor dude talked about the trailer I noticed that directly next to the path, at the point that appeared to be where the fire had started, there was a burned bicycle frame. It was white, whether from ash or prior paint jobs it was hard to tell. Someone had propped the bike like a statue on top of a blackened stump. People had been here already, it seemed. People may have been here when it started. Later there were rumors, always rumors, never conversations. The rumors said a person did it, I know who it was, the guy in the tent, the guy with the bike, the guys I saw on the trail, that guy's gotta go. The app that told me where fires were and how many emergency vehicles had been dispatched to them never circled back to talk about causation. It was engineered only to react.

Trees were still smoking. The ground, too. Nootka tugged but stayed obedient.

I regarded numbly the smoking hillside. One of the smoke pillars seemed more vigorous to me than the others. The distinct plume was about fifteen feet off the path and up a slight incline. There were trees nearby that had been downed by fire, and piles of ash where likely more vegetation had burned entirely. A few tan oaks and firs had been split in two by prior seasons' winter storms; their bottom halves remaining rooted and their tops splayed on the ground, burned. I leaned off the path and over the heat of the burn scar, trying to get a better view. The suspicious smoke began to balloon, growing thicker and whiter. It too was leaning, from a space I couldn't see between two dead trees, out toward the fire perimeter, reaching for an unburned bush. I think that's still on fire, I said. Numbness turned to adrenaline once more.

At the burn area the firefighters had left hoses around the perimeter of the scar, standard mop-up practice in case anything flared again. (*Mop-up* was yet another fire suppression term with disturbing military origins.) The neighbor climbed across the hose line to check it out, walking over smoking dirt and burned vegetation. Kelly and I looked at each other. Men are ridiculous, the look said. It wasn't remotely safe to be doing that so soon after a fire, and certainly not wearing sneakers.

The fresh smoke I'd seen was coming from a Douglas fir, a half tree, really—disease or circumstance must have weakened it long before the fire. Its bark had the usual rivets and wrinkles, but the trees' weakness was indicated by what it was missing. Lichen didn't adorn it; its branches were bare of leaves and its crown had fallen to the ground a long time ago. The neighbor dude reached the tree and jumped back again, startled. I repositioned myself on the path in a spot where I could see the burning dead tree. An ovoid gash in the trunk, about waist-high to a human, flickered like a portal. Inside, an apparition: a nest of flame. I had heard that in forest fires, sometimes a tree might appear to be whole on its outside while fire, having manifested within the bark, lurked within. A tree didn't have to be hollow for this to happen; there merely had to be enough oxygen inside the trunk for the fire to thrive. As had occurred at Big Basin, large trees could burn inside themselves for hours or weeks. It was one of many reasons wildfire areas were kept off-limits to the public in the immediate aftermath.

The fire glimmered freely inside the blackened tree, blurred by its own heat emissions. It was orange and blue and black. From time to time it whispered out of its peephole and shuddered into the open briefly; the smoke I'd seen was its advance scout. There was unburned vegetation a few feet away. It would only take a breeze to set these woods off again. Kelly and the neighbor and I didn't know what to do, so I took out my phone and called 911 and had the distinct displeasure of being the originator of the siren's wail.

The VFD rolled up quickly. I'd never seen a car on the path before; it was scandalous to see their big truck rolling through the woods. Three men greeted us. They didn't seem hurried. They didn't thank us for calling. They got to work. One climbed up the hill to the upper portion of the burn, to do another mop-up loop. Another picked up the perimeter hose, and the truck started pumping water to it. We pointed out the tree on fire, and the youngest man set to work on it with an axe. That's the only way to put it out, he explained, you have to cut down the tree.

It came down with a soft creaking noise. No thud, no boom. It was already dead. The fire, loosed from its fuel, quickly extinguished in the smoldering dirt.

I supposed I should have felt relieved—dodged a bullet, close call, what luck we were here—but I didn't. I felt forlorn about the tree being felled. Another one, dead.

Maybe it was the mood of the season, the spooks of the night before, but the fire inside the tree had appeared to me to be rooted, like the tree. The fire's entry point into existence was the same rich dirt I was standing on: the underworld, now surfaced inside a heart of wood. I had taken a picture of the heart fire and would later gaze at it repeatedly, trying to summon the feeling I'd had that morning on the path, the loss of something alive, and the strong conviction that something was being communicated to me. It had felt like the fire and the tree it burned had been in the middle of telling me something important, but I had caused it to be extinguished and the fire had been muted and sent home too soon. As I walked home on the path, my pain opened, like a wound rubbed raw from fear and relief and fear and relief. I also felt a strong sense of loss. I missed the tree's company, its evident poetry, already. I wanted to know what it had to say, what it had brought from the universes on the other side.

MEMORIES. In one, I am at a writing residency, alone in an A-frame cabin on an island, with the woodstove pumping and the wind from the strait careening up through the meadow, and from my window seat I see trees wave their hands at me, branches and leaves rustle, they are saying things, only some of which I will remember later because I decide to pull myself out of their reverie, afraid that if I listen too closely I might never return to regular ways of speech. They are fir and cypress and elm. Eagles sit in them. Rain knows their taste.

In one, it's a different era, a different forest, it's springtime at the little house on the hill. I am reading on the porch and I notice the Japanese maple that leans over the house is putting out soft, vermillion-colored leaves, which will soon size up and turn in the annual greening. Above my head every small leaf shakes as though wind-tossed. It's not windy. I look up at the tree's joints and limbs, and I realize they are buzzing. It's a literal, audible buzz, as though a cloud of noise is moving the parts of the tree. I stare at the cloud until my vision narrows. My eyes refocus: in the tree, there is a swarm of minuscule bees. They feed rapturously on the pollen of the new leaves, and their wings sing.

In one, it's cold, dry winter. I am walking on a recreation trail that threads between vineyards and estates on a disused railway bed, and I pause to rest on a bench. The view is all rolling hills, covered with grapevines. It's a regular walking route for me, a regular view for Sonoma, but today I notice that in the middle of all that agriculture, every few hundred yards, there is an ancient oak tree. The oak is so large it's become a geological feature: the rows of grapes part and curve around it. I can see the wires beneath the vines, small infrastructures of extraction contorting the plants into products. Amid such scaffoldings, the oaks appear to me chained. They crackle and root, but the fences and barbed wire hold them. I imagine the oaks untethered, the vineyards plowed under and replaced with medicine and food, and when I do this I can hear the trees cry for release.

In one, I'm in the hammock watching the gods dance, a breeze-dappled summer afternoon. I have recently read a book—Suzanne Simard's *Finding the Mother Tree,* that breeder analogy just can't quit—about how trees communicate with each other by sending information and nutrients through the networks of fungi beneath their roots. I eat four grams of psylocibin mushrooms. The redwoods speak clearly this time, using words, and what they say is, It doesn't even matter. This is before I have read of Fukuoka's similar vision, and the trees don't know who he is either. They say it over and over, and I laugh every time. It doesn't even matter. In the banality of my experimentation I easily comprehend what the trees mean: nothing humans do is real, and the trees don't care, and we are all here together in dirt. This feels to me somehow like the opposite of despair. To my human friends, it's boring when I tell it to them later.

AFTER THE FIRE on the path it was as though my senses had been readjusted. I had seen fearsome fire, in person. It had visited me where I live. It had taken the form of a tree's heart, burning. Fire had shown me a door to an underworld, then the door had closed. The gift it left me was the ability to continue inhabiting my home, my life.

My perception had been that fire was a relatively new presence in my life. The truth was that my prior life without fire had been exceptional. I happened to be born and live at a moment in which fire, while never truly gone, had been by force, law, and denial pushed—

suppressed—to the margins of human experience. But it belonged here, it had always been here, underground somewhere, and anyway, it would be back soon.

When I told people about the fire on the path I usually stressed how lucky we'd been, it could have been so much worse. So scary. Relative humidity. With closer friends, I talked about what it felt like to face a deep fear—the woods behind my house, on fire!—and have it turn out okay. But I didn't talk much about the tree burning inside itself. I didn't tell how this one, on that cool autumn morning, had talked to me on the path. I certainly didn't say I'd felt the draw of an underworld there in the burned woods. As with the story of my body, there would always be different ways to tell it. I could have said that a couple of acres of forest near my home burned that night. Or I could have said that the veil between worlds became too thin in this place, and I saw through it to something else, and it greeted me, but because of a chance combination of conditions it wasn't a world I was deemed to enter, not yet.

At home it smelled like camping. I felt adrenaline-rerun exhaustion kick in again. Kelly texted that she could hear the fire guys still mopping up in the woods behind her house.

I poured myself a tall gin and tonic and began to run a deep bath. The tub that came with the house was a behemoth. I could lie down fully inside it. The bliss inherent in this fact mitigated the drought-induced guilt I felt whenever I bathed lately. As the bathtub filled I grabbed a pair of scissors and went out to the garden. I brought back inside several long stalks of lavender and fragrant stems of rosemary and tossed them in the bath.

I lit candles; I poured salts; I placed next to the gin and tonic a glass of cold water and a small bowl with chocolate chips in it. I invited myself: Sink in and exhale. Feel the water. The way it ripples over skin, distorting the drawings of turmeric root and roses tattooed over my hysterectomy scars, the candlelight's reflection in the window that is not on fire.

I was trying to teach myself to tend to my body with as much care as I showed my garden. That afternoon in the bath, as the small fire no longer burned uphill from the small wooden house, when pain coursed through my systems I tried to notice it, not avoid. I tried instead to stay with it, to observe its route as though it were water, seeking low ground. I wanted to see if I could reinvent these webs of sensations as networks of connections. Perhaps, as Joanna Macy had suggested, my pain might be indicative of something more than a sloppy healing process. I already knew my pain connected me to the garden, and by extension to that patch of burned woods out there. I was beginning to suspect that if it was possible to stay within the trouble, then my conduit to doing so was going to be pain, not plants. I was going to need to become my own balm.

The bathtub was filled and I turned off the taps. I sank low, relished those submarine sensations. The scent of herbs and smoke wound through the steam, offering a kind of pleasure in the unraveling.

I took my hand underneath the water and began to rub my belly, the tea rose inked in its center. At the time I got the tattoo, I'd imagined this spot to be the origin point of my pain but in fact this was only one of its dawnings. I used the movements the massage therapist had shown me. She normally specialized in pregnant and postpartum women but had made an exception for me. Three strokes up toward the navel, three to the left, three to the right. Then a zigzag, downward, pressing hard from sternum to pubis. As I rubbed, I contemplated the actual shape and density of my belly, and I was mystified. What was in there? What had my body filled itself up with, what was the shape of its new power, once what had been was removed? I looked up. Framed within a skylight above the tub: trees as tall as sky. Today, a day of no devils, they were quiet.

A FUTURE MEMORY. It's May, the most ostentatious time of the year in a California flower garden. The roses are resplendent. I'm walking through the wild part of the yard with my friend Latria, a writer I've invited to stay for a couple of weeks so she can get some retreat time in before diving into another freelance assignment. She is round-cheeked and wears long purple braids. We're talking about the garden, I'm telling her how Max has done retreats here with the workers and I want to do the same thing for artist friends, a group of women writing together, imagine it. She is talking about this piece of land and its beauty, and at one point she refers to the house on the hill and the garden around it as your utopia here. She says this as a matter of fact. I stumble on it. Our utopia? There's that word again, yet another ideal historically linked with the lie of nature as a pure, wild, empty, other-than-human place. A utopia offers a false hope: escape. It is naïve. At best, I think, as we walk through the roses, utopia aspires to be the burned fruit orchard at Monan's Rill, struggling to keep going. And besides, it never works. I take pride in *not* fancying the wooden house on the hill as a homestead; we don't even grow vegetables. But Latria, a fourth-generation Black farmer from the South—a brilliant person who writes about land and its connection to her and her people, to labor and refuge and place—must be employing a more benign usage of the word utopia. She is not about to be lauding some remnant of Manifest Destiny. Maybe she means it like *idyll*.

You think so? I say to her.

Perhaps, I think, Latria is speaking more in the tradition of Ursula Le Guin, who in her essay "A Non-Euclidian View of California as a Cold Place to Be" defined utopia as an unmappable place, "such that if it is to come, it must exist already." A place of past, future, and becoming. I could agree with that. What was this hillside to me, if not a place that was always becoming?

This is the same springtime when, in a barn near Petaluma, I will dance at a quinceañera with Max and our friends Lina and Davida. Davida and Lina: organizers, musicians, actual woodland fairies. The party is at the home of Maria Salinas, an Indigenous Oaxacan activist we know through the local farmworker movement; it's her daughter's fifteenth birthday. Maria's husband, Bartolome, has taken what was a dilapidated barn outside the house they rent and built his own kind of utopia: a community event space. Twinkly lights are wound through the rafters; sheaves of dried marigolds hang from the eaves. On a wooden stage, a large band plays music with lyrics in Spanish, and people speak Chatino, an Indigenous Mexican language. By day, Maria holds Chatino language classes for children in this barn. At the quinceañera the table centerpieces are cowboy boots, the tacos and mescal are bottomless, and it's the first party I've been to in more than a year, since the pandemic began. My friends and I dance as though for the first time. In a break between bands, I go outside with Max. He smokes a joint while I inhale the night. Maria's garden glistens in the moonlight. Stalks of maize reach toward the constellations; the ground is a sea of squash blossoms. Beans try to climb up my legs.

They've built all this? I whisper to Max, amazed. Yeah, he says. It's magic, I say.

These are the kinds of memories I want to fill my home with. I want Kelly, her kid, and our friend Freddie to carve pumpkins at the picnic table in the backyard, and for Freddie to take a video using some kind of filter that makes us look like we're in an old horror film, all sped-up and grainy. I want a book release party, a birthday party, a retreat for Max's colleagues, band practice. I want workdays with the neighbors, clearing dry fuels from the hill. When the water runs out I want to replace the rosebushes with natives—clarkia, yarrow, western columbines, bush anemones—and see them flower in the heat. I want to make a batch of fresh chicken soup for Frank's frail wife after she gets home from a hospital stay, and months later to sit in a different hospital room with them all night as she transitions to the other side. There will be an evening when I squeeze Latria and Rahawa, also a writer, into my office, a former laundry closet that I've decked out in wine-red walls and a cushy armchair. I tell them as they browse my bookshelves how I made the shelves from redwood, how I covered the ugly white linoleum floors with faux Persian rugs. I want it, I tell them, to feel like the kind of room women didn't used to be allowed in. A not-Narnia in the redwoods. Like a mass of overgrown greenery you push through to discover a small oasis, created with care: a secret garden.

ALL THESE UTOPIAS Max and I were making had already been made, even if I wouldn't choose that word for them. In the bathtub that day, after I met fire in the forest, I let the flowers I'd put in the water with me stick to my body. That was one way of staying here, I supposed: to make it magic, until we had to leave. This was a kind of power pain had given me: to make it beautiful even when it hurt.

November

· 11 ·

Owl

On Election Day the skies were clear. The tarot card for the day was the Hanged Man: Unknowable outcomes. Patience. If you want enlightenment, you've gotta go through some shit first.

Near the house where I lived, the path fire remained extinguished. The veil had been drawn closed. The previous night, Kelly and I had volleyed our anxieties over text messages into the small hours—the election, continued fire weather—and I took a painkiller in an effort to sleep deeply, but I still woke up too early. I was restless with anticipation, so I went to the coast.

I was the only human there. Three small deer grazed near the trail. Rattlesnake grass muttered quietly as the wind shook its dry seed pods until they formed waves on the landscape. It was cold enough to wear a hoodie and sunny enough to need a baseball cap, and it was still so dry. The trail turned to boardwalk where it passed through areas of sensitive wildlife habitat, but where it was dirt, the ground was cracked underfoot. In wetter months, burrowing animals were often seen hard at work here—there must have been entire cities of tunnels in these cliffs, telltale dirt piles denoting the networks of ground squirrels, gophers, weasels, and whoever else lived down there.

Today the ocean was ill at rest; choppy surf provided a steady white noise. My pain was background chatter, feedback. As I walked I worried through election outcomes in my mind. I was sure the fascist would win. I rounded a curve in the dirt section of the path and approached a spot where there were two large, foot-wide holes in the ground. As I sidestepped them I noticed there was something brown inside one of them. It looked like an animal. And not like a gopher. I slowed, startled, but continued my approach. It was a very small owl. In daytime. Sitting down. In a hole. The owl had perfectly round eyes with large pupils, and from its low hunch I could see intricate variegations in its crest. It looked like a creature from another dimension, a mythical animal—a bird of the air who lived underground. The owl, also startled, didn't withdraw. It stared at me, and I off-roaded into the brush in order to give it space.

A few minutes later, on a different curve of the path where it bordered the cliffside, another hole. Another owl. This one was standing up like a person. It had unusually long legs for an owl, which gave it an almost goofy stance. It had no visible ears, but like most owls, its feather patterns gave it the appearance of a humanlike face. It was the oddest little critter I'd ever seen.

The owl and I stared at each other for what felt like many seconds. Then it took off. Its long legs collapsed out of view and it sped away quickly at a low angle above the fescue. It was difficult to distinguish the owl's airborne shape against the background of the sunbaked grass, but the white patterning in its plumage gave it away. In flight the owl looked larger than it was. On the ground it had appeared awkward; in the air it looked more like a predator. The menacing grace of a raptor in flight.

As soon as I got home I emailed Stephanie, the family bird expert, and she informed me that I had experienced a rare sighting: a pair of burrowing owls, *Athene cunicularia,* of the genus *Athene,* as in ancient Greece, as in democracy.

In recent decades burrowing owls had become threatened in various areas of the Americas, and in California they were listed as a species of concern, but still not very well protected. In Sonoma County they'd been in decline since the turn of the millennium. The owls' habitat was grassland—they nested in the ground in pairs, usually in holes abandoned by other burrowing animals, and they often returned to the same spot year after year. Their numbers had declined because the grasslands they nested in were being developed, often to build golf courses or exurban housing. In their own way, burrowing owls were also residents of the WUI.

I skipped the mammoth rocks today but did my usual pilgrimage to the water's edge to gaze at a certain boulder I admired. It was gargantuan and situated several hundred yards offshore. On its slope, blanched white by sun and guano, many dozens of birds sat. I loved the rock because it seemed so fixed to me, it was enormous and impenetrable. I often thought about eras when it had been connected to the shore, when the cliff extended out that far, and I wondered if birds liked it then too. I took a picture of the rock, which joined the hundreds more pictures of the same rock that had accumulated in my phone over the years. I imagined a distant future in which I would finally line up all the photos in chronological order: a manual time lapse.

I looped around and walked back to my car. The owl watched me the whole way from a perch on a low-lying bush. I decided to take the birds as omen: Athena would win the day. The fascist lost the election. There was no insurrection, at least not at that time. I didn't know what the results meant for the long-term future of the country, but for me it meant an end to solitude. Max was coming home.

Would you ever make a THANK YOU FIREFIGHTERS sign for our house? I asked Max.

Absolutely, he said. But mine would say THANK YOU *VOLUNTEER* FIREFIGHTERS.

Our volunteer fire department consisted of about ten people, all of whom had other full-time jobs. They worked hard.

How about a hashtag sign? I said.

A what?

#WESTCOUNTYSTRONG, or whatever.

Absolutely not.

We were driving in the country outside the town of Healdsburg, which the Walbridge Fire had almost burned into a couple of months earlier and the Kincade Fire had almost burned into the prior year. The signs were now a year-round fixture: thank-you messages in children's handwriting accompanied by cute stick-figure drawings of houses. Big, bold marker in all caps scrawled on cardboard:

THANK YOU FIRST RESPONDERS!

#SONOMASTRONG

#HOPEWINS

In recent years the language of survival had taken on a strange homogeneity. After any disaster or terror incident, the signs went up. The formula was usually hashtag + town + strong. On social media, hashtags were often used to practical effect by disaster personnel, governmental and otherwise, as a way to identify breaking news and information amid nonlinear timelines. But civilian, offline use of handwritten hashtags unsettled me. It framed survival as a consumer experience. Hashtags were for brands or trends. Whether organic or optimized, a hashtag was a thing intended to be followed, the words it prefaced marketed as desirable, packaged for consumption. As though a person should want to have an occasion to be *hashtag strong*.

Is that really what we're going for, continued Max, in the face of climate change? Being strong? That's all we've got?

I agreed: it wasn't only the hashtag that was the problem. Fire especially, of all so-called natural disasters, seemed to inspire a fetish for strength. Such declarations of strength felt to me like a desperate assertion, not a show of solidarity.

I mean, Max said, someone is printing up those stickers and signs. People are buying them. This is how capitalism gobbles itself up! Like, let's take the act of surviving an event that is the direct result of the ravages of capitalism and make it into a slogan?

Wildfire, as an extreme weather–driven event, was a literal demonstration of humanity's weakness, both acutely and in the big picture. I understood that the hashtags espoused a metaphorical strength, that people needed something to keep going emotionally, and I didn't know what it was like to go through what a lot of these people had gone through, this close to catastrophe. But *strength* implied rigidity, which meant not changing. Rarely light on their feet, the strong (with the exception of Muhammad Ali, as my dad often reminded me).

It's like how people talk about cancer, I said. Like it's a battle, we'll fight it, we'll win.

Max shifted his posture in the driver's seat. His mother was currently being treated for late-stage lung cancer. People lost the battle to cancer all the time, and that didn't make them any less strong or their deaths any less tragic. Susan Sontag spun in her grave again.

The road wound through alternating patches of sun and shade: grapes, oak, grapes, oak. Hashtag hope, the signs read.

In my experience, any discussion of the realities of climate change, in any locale—fire season, hurricane country, places still reeling from millennia of resource exploitation—inevitably ended in an invocation of hope. I never knew what to say when people talked about hope. It wasn't a concept that resonated with me. As much as I craved a future I could look forward to, the version of reality in which communities rebounded, strongly, time after time from disaster after disaster was not exactly a hopeful one. The prospect seemed grim.

Max, do you think that I am hopeless? I asked. I was definitely the pessimist in our pairing, and I was aware it sometimes rankled him.

Naw, there's still hope for you yet, he said.

Haha. I mean, like, a person who does not have hope. I don't know if I have hope.

He considered it.

I think, he said, that hope is maybe the wrong word for right now.

But I do feel hopeless about the fires, I said. About the climate.

I know, he said. But I don't actually think you're a person without hope. It's more that instead of just hoping, you understand that we need to be engaging with what's going on. I guess the problem with hope as a goal is that it usually stops there. *Hope,* the way people use it now, implies a return to normalcy, like a fixed solution, and that usually translates into doing the same shit as always, but expecting different results.

This was also said to be a definition of insanity, and it was a favorite truism of Max's grandfather.

Max was an organizer; he spent every day with working-class people who had wild hope, the kind of hope that drove mass social change. The hope Max saw with the workers was different from hashtag hope. It was more of a belief that things could change, that people could change them, because it *had* to change. Over and over, he had witnessed people with little reason to hope—people brutalized by economics, power dynamics, inherited trauma—challenge the most powerful structures and people in their lives. Bosses. Politicians. Entire institutions.

That kind of hope, Max continued, isn't about replicating some imagined normalcy. It's about trying to change things, even though you know you might not win.

My own recent dalliances with hope had led to failure. As a person who lived in an ecosystem that was in the process of self-immolating, I found hope to be even more rigid a standard than strength. It felt fake. In facing the fact of climate chaos in my own life, I experienced grief, determination, awe, and a strong sense of continuing on, a desire for change. But hope wasn't a guiding aspect of my lived existence. As a person who lived in an injured body, I had hoped I would be healed. I hoped that the next treatment or surgery would stop the pain. I sometimes even hoped for a shocking diagnosis, a big bad explanation for what was *really* wrong with me, the deliverance of a type of hope that provided a clear solution with a charted course to recovery. For me the thought of hope summoned the deep crash of disappointment when it didn't happen, when I couldn't fix it, when everything was not, after all, okay. Every time my next big hope fell through, I was ashamed of having believed it to be possible. It was embarrassing, to have failed at a happy ending, whether that meant a pain-free body or a more just society. How mortifying it was, to be unable to avoid the innate hubris of my species.

I fiddled with the car radio. Classic rock again. There must be a way of moving forward in chaos that did not require fealty to a long-exhausted cliché.

A couple of weeks before his writing studio burned down and a few months before he died, Barry Lopez wrote in *Orion* magazine that he had begun to back away from the false promise of hope in the era of climate change. He advocated shifting from optimistic denial to a kind of loving realism. "If we are to manage the havoc," he wrote, ". . . we have to reimagine what it means to live lives that matter, or we will only continue to push on with the unwarranted hope that things will work out." The only way to do this was to stop pretending the future would be rosy and to "embrace fearlessly the burning world."

A different article I had read in 2014, when the weather had been noticeably changing but disasters had still appeared to be occasional, had often come to mind in the years since. In a short interview in a now-shuttered magazine, environmental journalist Brooke Jarvis had interviewed Chris Jordan, an artist whose work focused on human waste and the environment. Jordan had shot a photography project on Midway Island in the North Pacific, where albatrosses breed, and where human pollution caused baby albatrosses to die en masse. While nesting, the juvenile birds ate only what their parents brought them, which was increasingly plastic trash from the ocean, and so they'd often starve to death before they fledged, their corpses rotting and leaving only bones and feathers in a haunting circle around the nonbiodegradable contents of their stomachs. These remains were what Jordan photographed. In the article, he spoke of how depressing it had been to tour schools and communities and show these photos. Everywhere he went, after his presentation he was approached by audience members in tears, distraught. What could they do about this tragedy? they asked him. How could they find in his images some sense of hope?

"That's not my job," Jordan said to Jarvis. His job as a photographer was to witness, and then to make something that might help his viewers sit with the truth of what was happening. In this way, Jordan's artwork was in part an exercise in grief, solastalgia in practice. Only by really sitting with the reality of these birds' lives—the horrible death and species decline, yes, but also the other wonderful things he'd seen on Midway, the wild beauty in their domestic lives and sweeping flight arcs—could people then proceed toward something resembling hope. In the interview, Jordan called the feeling he had observing the birds—the same feeling he was hoping to inspire in viewers—an "ecstatic sadness."

Max and I were now within the city limits of Healdsburg, a former agricultural town that had in recent decades been reinvented as a wine country destination. The car inched in townie traffic past tasting rooms and luxury hotels. A new construction project was going up: luxury condos on the site of an old mill. There was a sign with a computer-generated picture of light-filled, high-ceilinged lofts surrounding a courtyard. In the picture, the sun was shining and white people did leisure activities. The stoplight turned green and as we accelerated I caught the ad's tagline: #WINECOUNTRYLIVING.

I don't think you're hopeless at all, Max continued. But that word is kind of played out. I think the real word we're looking for, even though it's loaded for a lot of people, is faith.

I knew what he meant, but since my own recovery had ended in my not being recovered, I had been wary of that word, too. Concepts like hope and faith had little to do with my experience of living inside a damaged body on a damaged planet. My experiences of harm and renewal were ever evolving, my body perpetually extending and then withdrawing the promise of a third-act resolution. Since the night of the lightning, I had become much more interested in the vulnerable chaos of uncertainty—the messy perpetual present tense that my therapist encouraged me to think of as being *with* pain, instead of being in pain. Being with pain was like living with fire in that sometimes it propelled me into new awareness and connection; sometimes it sucked. Both manners of living with uncontrollable events required me to let go of any sense of an ending. My experience of living with—and within—chaotic natural bodies proved to me the lie of a fixed destination. Since the early days of my injuries I had learned that gunning hard for specific outcomes was a bad way to live. Technically, the word hope simply meant expectation. It didn't necessitate salvation.

Max took his hands off the steering wheel and gestured to the burned hills.

Something has happened here, he said.

He meant wildfires, climate change, looming sixth extinction, all that.

And, he continued, hashtag hope cheapens the complexity of what has happened. It cheapens the care we need to show for ourselves and for nature now. The problem with that kind of hope, he said, is that it still imagines that humans are in control.

I thought of my experiments with control and faith in the garden. I had initially been tempted to see my experience of finding renewal in gardening while I was in recovery from medical harm as a redemptive arc. Cue music: *When her own capability to reproduce has been cut short, a woman finds comfort in nurturing plants, bringing forth new life into this difficult world to heal herself—and maybe, just maybe, unlock the key to healing the planet in the process.* The emotional tenor of such a pitch was one of hope, and it was broadly applicable to my experience: I did feel the high of creation when I was in a garden. When a rose blossom first asserted itself and then dozens of its kind followed, I became giddy with their scent, that adorable pink. Renewal! Semper virens! #SELFCARE! Although I found the act of gardening to be revitalizing, my garden was not the whole world; the sweet scent of my Jude the Obscure rosebush wasn't going to change the fact that one day, it would burn. Gardening was less an embodiment of hope or faith and more of an object lesson in understanding that every action—and the idea of forward progress itself—was complicated. You cut a thing, you fed a thing, there was water, beauty, sweet, sweet fruit. Sometimes it went wrong. Mostly it didn't. There was no redemption here, only an ongoing act of livingness, a refusal to stop tending to a life.

West from Healdsburg, the road meandered through a neighborhood of houses that were tucked like afterthoughts onto a woodland hillside. It was oak: live, black, white, and all the other kinds of oak trees that to me were still indistinguishable. The land below and between the trees was covered in dry grass. As the road got twistier I focused on the minutiae of the scenery, looking for signs of the recent fire. There were none, unless I counted the prevalence of luggage racks atop cars in driveways, evacuation-ready. There were also rather a lot of FOR SALE signs. My jaw clicked in its perpetually self-defeating effort to channel nervous energy away from my brain. Today my pain felt like a rabbit, nervous and fast. Over a gap-toothed wooden bridge and down a dirt road we landed at the bottom of a deep gulch, and we were in the preserve. I rolled up the window to the dust. The road ended here, morphing into a large dirt path.

We were visiting four hundred acres of mixed oak woodlands in a gulch between short hills on the eastern side of what had been the Walbridge Fire, which had carried itself here from its western origins near the woods where I lived, a thirty-minute drive away. Max and I were here to volunteer with a local environmental nonprofit that bought large tracts of land for conservation. Two of their preserves had been partially burned in the Walbridge Fire, and we were entering one of them that day in order to help the land recover. The real reason I was here was to stand in a burned place and touch it. I was done with peeping from roadsides, imagining the destruction, or extrapolating the impact of large, destructive wildfires by observing smaller burns like the one on the path near my home. I wanted to see what fire could do.

We parked and gathered in a loose, socially distanced circle with a couple of staffers and the other volunteers. There were liability waivers to sign, and we did so, balancing the forms on the trunks of our cars. The stewardship manager oriented us and passed out hand tools. There were about twelve people here, all white, except the stewardship lead, who appeared to be Latinx. Most of the volunteers appeared to be over the age of fifty. We wore a variety of sun hats and cloth masks, for Covid and also for dust. Our water bottles stood lined up in the dirt by our feet, shouting their like-minded slogans at the sun.

I was standing on the line where the fire had been turned back by humans. To my west the hills rose up in a panorama of char black and toast brown. The fire hadn't burned its hottest here, which in part accounted for our presence. It was safe enough to bring volunteers in and not get in trouble with your insurance company. Deeper in, the fire had been far more destructive, and in those places vegetation and roots were still smoldering, dead and damaged trees were still in danger of falling, and roads and bridges were not passable. I was still being kept to the outskirts of the burned places, but for today it was enough.

Looking around, it was difficult for me to determine which trees had been burned in the fire and which were simply heat-shocked, a phenomenon in which a tree's leaves dry out suddenly and entirely when the heat of the fire overtakes them. One of the staffers said it can happen within a span of seconds. A lot of these top-killed oaks probably had intact root systems belowground; they would live, we were told, and it would take a year or two for them to recover and begin growing again. Most oaks (there were hundreds of types native to California, and dozens in the greater Bay Area) were well adapted to fire. Their thick leaves and bark helped resist flames; their large roots stored carbohydrates that would feed new basal sprouts after a tree was top-killed by fire. And acorns were a favorite snack of many woodland critters, who stashed them in various locales on the ground and in other trees and thereby distributed a seed bank that could become new oaks if old ones were killed.

The staffers had already been here for a few days, clearing burned brush and hazardous trees. The volunteers' task was to mitigate potential erosion and nutrient loss caused by the fire and prep the soil for rainy season. It was early November. Rain was predicted. Nobody I knew celebrated this with the enthusiasm it merited during the worst fire year yet on record in California, at least not out loud. The ocean was a complicated mover of weather and it was common in this area for rain to appear on ten-day forecasts then disappear. The prospect of an absolvent rain remained largely unspoken, but it floated beneath the surface of conversations, lifting our group's volunteerism with an expectant energy.

The first thing the staffers told us was that the damage we were there to repair was the result of the firefight, not the fire. When the suppression teams had created their successful fire line here, they'd used the natural contours of the land as a starting place: the path, which in turn followed the same courses as winter storm runoff, and a dry streambed. They had widened these natural barriers into larger fire lines using bulldozers. In doing so, they had moved a lot of soil, inadvertently filling the creek with it. Dozer work could be dangerous: I saw a pair of tractor tire marks going up the hill above the creek on a grade that was practically perpendicular. One had to admire the extreme trajectory the dozer had taken up the hillside. Acrobatic machinery. Where I stood, on the preexisting fire road, more tire tracks had made deep grooves, displacing mounds of dry, loose topsoil. These disturbances to the soil would be ideal territory for invasive species like yellow star thistle, an exuberant and rapidly propagating plant that shaded out other natives and was a fire risk. Once the wet season began, the dozer tracks also had the capacity to change the natural downflow of water in ways that would allow it to strip the soil of nutrients and erode the structural integrity of the hillside.

There were many places within this preserve in which the greatest harm had been done by the fire, not the humans. Soil was made up of a combination of water, air, organic matter—which could consist of plant or animal matter and micro- or macroorganisms—and minerals in the form of broken-down rocks. Fire had the capacity to release many nutrients that improved soil's properties, including nitrogen, phosphorus, and potassium (the same *NPK* content that was displayed on the labels of garden-scale fertilizer I bought at my local nursery). But extreme fire changed the physical behavior of soil, making it less able to absorb water, which like the dozer tracks led to such negative effects in the landscape as erosion, sedimentation, and changes in water runoff patterns. In a wildfire, species and processes were interrupted or halted; soil nutrients were burned. In some places, the right confluence of these circumstances led to abundance; in others, nothing sprouted after a fire except invasives that bogarted nutrients and starved out other species. The inbuilt and human-assisted regenerative cycles of California's ecologies had been diminished for a long time.

Max was put to work shoveling piles of displaced dirt out of the creek bed. I joined the senior citizens in less physically demanding work. Our task was to smooth out the dozer lines, using our tools to even the soil and ideally approximate the pre-bulldozer slant of the hill. One of the staffers had a Pulaski, a sort of combination axe and pickax. Pulaskis were named after a U.S. Forest Service ranger from Idaho who in the 1910 Big Burn had saved the lives of sixty of his fellow firefighters with quick thinking, the use of an old mining cave as a shelter, and some wet blankets. These were wildland firefighting tools, and we were using them to mop up the damage the fight had done.

I found that the best tool for reintegrating the dirt into itself was a McLeod, pronounced like *cloud,* which was part shovel and part large-tined rake. I set about hacking at dirt clods and channels, then raking them smooth. The smell of dirt and charcoal penetrated my mask. In the dirt I found granola bar wrappers; emptied packets of protein goo like the kind used by athletes and mountaineers; beef jerky packaging; plastic water bottles; random bits of wire; discarded surgical masks; and a single mateless work glove. These were the residual costs of containing a wildfire: trees felled by tractors, undergrowth uprooted in haste, soil riven by tire tracks, a creek that became a dirt mound. Trash. Human stuff. Someone joked that we were raking the forest—a dig at the soon-to-be former president, who had belatedly addressed the wildfire crisis publicly but had done so with his usual counterfactual assertions. The work felt to me more like sweeping a desert. Was this still a forest? It still looked like one from afar. Up close I wasn't sure. The hillside was sapped of any hint of life.

Leveling the dozer lines was simple and sweaty work, but I found enjoyment in it. I always took satisfaction in busywork. I had been a great receptionist in my early twenties thanks in part to my love of the meditative aspects of collating photocopied documents. I leaned and thrust my tool forward, past my arms' natural reach. My back bent and stretched forward too, and a warning flare ruffled somewhere beneath my buttocks. My thighs tensed in compensation, trying to take on the workload of my injured parts. I kept going. Hashtag tough cookie.

That day, it took a dozen people six hours of handwork to mitigate about a hundred yards of dozer lines. After I finished my work, I stood under a heat-shocked oak tree and looked east, where the Mayacamas Mountains were black and craggy from this and prior years' fires. I tried to imagine the layers of labor that were needed to ensure this minuscule part of California's landscape could continue to thrive. What was the plan here? A group of middle-class volunteers like myself could rake this forest and then go home and feel good about it, but were we helping? Or were we merely reenacting the theater of conservation, trying to make the land pure after fire had sullied it. No matter what it signified, this seemed to me an insufficient amount of labor. Fire mitigation, preventative vegetation management, restorative agriculture practices, home hardening. How much labor would it take to change the course of events on this land? How would those laborers be paid, housed, and kept safe? Machines and algorithms couldn't do this type of work. It was intimate and handsy, and it entailed judgment. It was work that required an eye for and knowledge of the landscape; it necessitated care. It had to be people. It had to be us.

· 12 ·

Oak

A future memory. When she sets fire to a forest, Sasha Berleman smiles. It's hard to tell she's smiling, because her face is almost entirely obscured by a pair of giant aviator sunglasses, and her pale skin is covered in black charcoal streaks. She holds a one-and-a-half-gallon drip torch, a canister that looks a bit like an oversized version of an old-fashioned oilcan. The canister contains a mix of gas and diesel fuel, which drips out through a looped spout to a wick. When the fuel is ignited, an oxygen inlet valve feeds it, and the loop in the torch's neck prevents fire from reversing back into the canister. Sasha dips the torch toward the forest floor. A millisecond later, small ripples of heat hover over the earth. Her aviators reflect the sky. She's having fun. She takes a moment to observe the flames, then goes back to her walkie-talkie, over which crackle the voices of about a dozen of her peers.

We are only a few miles from where I raked dozer lines that November morning in 2020. It's a year and a half later, springtime, on the heels of a wet season that was not nearly wet enough. Today it is not too dry and not too moist, and the wind is just right. It's a burn window, and Sasha and her crew are taking full advantage. We're standing on a hill at the edge of a school parking lot. Downslope from us is a tract of oak woodland. Below that, the vastness of vineyards, conspicuously green against the native landscape. It is a small burn, only ten acres today, but a strategic one. In addition to mitigating fire danger for the school, the burn helps create a strategic firebreak between the

Mayacamas foothills and the city of Healdsburg. This oak woodland is, as one Healdsburg firefighter tells me, the last line of defense.

It is morning. Foggy. The burners and observers huddle in a circle. I'm an observer, one of a handful of civilians being babysat by an intern, who has just reprimanded a woman in a bright orange puffer jacket for not wearing fire-safe natural fibers as instructed. We are all briefed on the fire plan, the map of the area, safety protocol, and the weather, which will be monitored and reported every hour. Then without fanfare it begins: two teams of fire practitioners trudge in military formation past me toward the trees. Each carries a hand tool and their own food and drinking water for the day; some brandish drip torches, while others hold manual spray pumps attached to backpacks full of water. The observers are mostly white folks, as are the fire officials, folks from Cal Fire as well as local fire departments. The burners are a mix of white and Latinx people, mostly younger than fifty, many of them recent graduates of Fire Forward's training programs. Their clothing is fireproof, industrial yellow, and heavy; I know that some of them are women, although identities are indeterminate beneath all that gear.

This isn't a cultural burn: from what I can see, there is no land acknowledgment or mention of traditional ecological knowledge at the initial huddle, which surprises me. As an observer I haven't been privy to every part of the crew's prep meetings, and I know that Fire Forward does often try to seek local Indigenous collaboration for their burns. But this morning the cultures I see represented are mostly those of fire suppression. Fire trucks and acronyms are everywhere I look. Sasha's colleagues, along with members of a volunteer-led community burn program called the Good Fire Alliance, are here at the behest of city and state fire agencies, who are in charge and whose protocol the burn crews must follow. As I watch them line up I think about how, since the 2017 firestorms, I have witnessed more popular acceptance of the need for good fire. When it comes to land, sometimes government agencies, Indigenous communities, nonprofit organizations, and fire suppression crews have goals in common, but often they don't. The distinctions among different cultures of good fire are porous and ever evolving. As good fire becomes mainstream again, I'm curious to see how they'll continue to interact. When I interviewed Margo Robbins, the Yurok fire expert, she told me she was excited that more people outside Native communities are starting to realize the land needs fire. Different people have different opinions about this, she said, and that's their right, but I think it takes all of us. Native people can't do it on our own, she said—we don't own that much of the land, for one thing.

Whether acknowledged or not, the burn today proceeds like a ritual. The fire practitioners start in a line at the edge of the pavement, on a road at the top of the burn unit. They will work downward and inward on a steep patch of oak forest. Fire naturally moves uphill; by using that behavior to their advantage and reversing its direction, they can ensure the fire will be smaller and more containable. Each practitioner angles a drip torch toward the grass; instantly, a ripple of warmth can be discerned in the air above the ground. It takes another moment to recognize flame but soon fire is evident. The burners move in a line across the land, repeating their actions, and it grows from there. Drip, light, drip, drip.

A good fire starts small and quiet. Neat rows of flames vein along the ground. When the fire collides with ideal conditions, it blooms bigger and deeper. Within a few minutes, I see flames combust small bushes on the ground and ascend toward lower branches of oak trees: the fire grows taller; as high as my waist, then higher. White smoke blows evenly. There is a perfect slow wind above our heads. On the burning hillside the air moves up and the wind moves down, and their collision creates vortices of oak leaves, mini fire-tornadoes that gather everything from the forest floor and heft it alight. It is easy to see how such a vortex might grow bigger, might plow across rivers and highways, or swell upward into a cloud that creates its own weather.

I am still afraid of fire, but I'm also increasingly curious. As the fire grows, I try to use my senses to experience it without imposing any narrative or moral value on what I'm seeing. Sight and sensation: fuel, air, heat, and the ripples of those elemental reactions. Sound: fire is loud. I can hear a bright, unsteady cracking sound coming from deeper in the oaks. Smell: smoke. My body reacts habitually by recoiling; I push it to keep feeling. There is a faint warmth emanating from the forest. Birds circle overhead. In the meadow bordering the oaks, voles scamper from one hole to another. Outside the tree line, a pair of towhees chirps and scrapes at the ground, presumably looking for bugs fleeing the fire. On the hill above vineyards and corralled oaks, eagles swoop. Swallows scatter. A junco hops around erratically in the grass near a burned patch. Displacement behavior.

Now the fire grows bigger. It burns into the interior of the acreage, where I can't fully see. On the perimeter, where I am, it spreads slowly along the roadside. Occasionally flames orbit themselves in a spiral on the ground, combusting duff and dried leaves. I observe, alarmed at the speed with which fire overtakes a bay laurel, and the oily tree combusts rapidly—*whooooosh,* a verdant blowtorch, sudden excitement. The madrones, too, burn hot, but only the foliage; the trunks stay naked and alive. The fire plows through ground litter, and when there is nothing but dirt left on the ground, it burns upward. A few times flames rise into the branches of a larger oak. When this happens, the lower tiers of branches burgeon into flame with alarming speed. They burn. But then, they go out. The canopy doesn't catch. I don't know why. Fire-adapted, the intern says. It seems like alchemy to me.

Observing the fire, I am mesmerized, as anyone who's ever spent hours around a campfire has been. But this is far more exciting than an evening by the fireplace. The fire is dynamic, it's on the move, and the people are moving with it, tending it, following it, understanding it. There is something about this fire that feels so familiar to me. Not because of the wildfires; it's more like an early memory of other instinctive interactions I've had with fire.

One thing I hadn't expected: Fire is exuberant. It's joyous. It dances. I can see why people joke that all firefighters are secret pyros. It's so much fun.

As children, my brother and I read a lot of Ray Bradbury books. I was captivated by Bradbury's visions of other worlds and other futures, and later I came to appreciate his sharp sentences too. There was little frillery with Ray. At the end of his novel *Fahrenheit 451*, there's a scene with a campfire. The protagonist, a former fireman, is on the run from the fascist government. In this world, the job of firemen is to burn books. There are no wildfires or house fires, except the ones the firemen set. When our hero refuses to burn books, he has to flee. After his escape from the city he falls in with a group of hoboes, exiled former professors, whom he initially mistrusts. They light a campfire to warm themselves in the cold night. He watches the fire, at first afraid, then captivated. He hadn't known fire could be a nourishing thing and not a bringer of destruction. He hadn't known it could smell so good.

Now the wind changes. The smell changes. A strong draft of smoke blows over to where I stand. I see it coming and begin to move in a different direction. I walk briskly, not wanting to display too much alarm. The smoke is opaque and smells like every bad thing that has ever happened to me. Sense memories kick in, and here's that familiar anxiety whorl: the sensation of my body at risk. I walk faster, the smoke grows leaden. My stomach flips and I feel like vomiting. Spatial depth vanishes in the span of a gust. When I realize the smoke is so dense that I can't see where I'm going, I panic a little. My senses constrict. I'm underwater but there's no water. I can hear a few people talking across the parking lot, they sound very far away. The auditory crackle of the fire, still about a hundred yards away, is muted by the smoke. I start jogging, I'm not in danger of being actually lost but smoke comes faster, thicker, with a yellow glow. Now comes the heat. All those summer suns. It is terrible. My lungs and heart close tight.

I run the last few yards out of the smoke cloud. Nobody around me—a mix of firefighters, Fire Forward support staff, and other observers—thinks I am overreacting. They feel that heat too. They know fire is always dangerous, even when it is under control. I take my inhaler from my pocket and puff, and I hear a jabbering of the intern's walkie-talkie as a man's voice explains the smoke onslaught: a couple of Douglas firs in the middle of the acreage caught and were fully engulfed. That got exciting for a minute, the fuzzy voice jokes. It's okay, the intern assures me. Those ones are supposed to burn, he says. The smoke settles back into a predictable pattern and the burn goes on. The interior of the forest is hot now, truly on fire; it doesn't look to me like a controlled anything. It looks wild.

I turn again to the fire. I focus on a small flame, one that is blooming on some low grass on the edge of the woods. At first, I don't see fire here; the dirt is dirt. But as I watch the ground I soon see that the dirt is blue, then orange, then it is a flame. The fire appears to me to erupt from the earth. It tremors in the oxygen resting on the surface of the ground. The fire and the soil are so obviously, physically connected.

I later describe this experience to José Luis Duce Aragüés, a prescribed fire training specialist from Spain who works with Andrea Bustos, the Ecuadorian prescribed-fire practitioner, at the Watershed Research & Training Center, a good fire–focused organization much farther north in California. José Luis is in my kitchen when I tell him this story. He and Andrea are crashing at our house after leading a training in Sonoma. For us José Luis is cooking a real tortilla, the traditional egg and potato dish of the Iberian Peninsula. He is using the largest cast-iron pan we have and a dozen eggs. The tortilla is so heavy that José Luis, a fireman, a person made of muscle and seemingly no fat, has trouble flipping it. After we eat—it's delicious—I tell him the story of how I saw the fire come from dirt, and how this moment at the prescribed burn had made me realize that fire is part of Earth. José Luis speaks with a Castilian accent, and his eyes are deeply set into his flame-tanned face. He says, Yes, this is how it is, with fire. Fire will teach you things you thought you already knew about the land.

Then he tells me his own story of epiphanic relations with flames. Both José Luis and Andrea have traveled the world making good fire. Once, José Luis was doing a prescribed burn in El Cerrado, the grand savannah of eastern Brazil. As soon as he arrived there, he tells me, he sensed a sort of mysticism to the scenery. Fire belongs to that landscape, he says. And you can feel it. So one afternoon they were burning. José Luis was alone in the middle of the unit, which is the term for the designated area of a prescribed burn. Things were moving nicely. Then he experienced something indescribable. There was a wall of fire, he recalls, and it was coming slowly toward him. This was a head fire, the type of fire that moves with the wind and the slope, but it didn't act like a head fire; head fires are hot, large, and fast. This unusual fire approached José Luis slowly. It didn't feel hot. When he moved toward it, it moved back a little bit. He looked at it, it moved forward, he moved again, and they went on like this. The only way he could describe the feeling he had in that moment, José Luis tells me, was as *intimacy*. There was connection. It was as though the fire was allowing him to be close to it. The moment of intimacy felt long but probably wasn't. He heard someone yelling, José, you have to get out of there! He stopped one last time and looked at the fire and he said, Okay, and moved slowly away. Now, he says, he feels fire in a different way, because there's this relationship.

The way José Luis speaks about fire sounds a bit like the way I talk about plants.

I ask Andrea and José Luis, both experts in the science of fire, if they think fire is alive, and they both reply instantly: Oh, yes.

José Luis says that he feels fire's personhood when fire responds to his touch. When you're using a drip torch there's a way, he explains, that you can grab fire. You can tell fire: Okay, I'm here. I'm bringing you over here, come with me. Like a shepherd. And it chooses to come.

Of course this is very evident with wildfire, Andrea adds, because wildfire looks so alive. But prescribed fire is also alive. You might think you control prescribed fire, but fire will make their own environment. They build their own environment. They decide.

It's after dinner now, and José Luis and I are on the couch in my living room; Andrea is sitting cross-legged on an armchair. We're all drinking tea. Andrea seems to disappear inside her large hoodie. Throughout the evening she speaks quietly, but with intensity of purpose, and always with eye contact. When Andrea talks to me about fire, she uses the pronoun they, and I notice that José Luis often refers to fire as she. I imagine part of this is the language difference between us: Spanish is their first language, and we are talking in English. In Spanish, however, the word incendio—fire—is a masculine noun. I point this out, mostly as a humorous aside, but it's an interesting question. If fire is alive, does it have a gender?

José Luis considers. He probably finds himself referring to fire as she because fire helps to give life. A creator, feminine. Andrea offers the word for fire in Quichua, an Indigenous language of the Andes: *nina*. Like a girl. Otherwise, Andrea says, smiling, fire is definitely they. No binary. They are a unique creature.

In my first few years of wildfires, I had related to fire as a destructive force, the same way I thought of my body's pain. I had hoped that perhaps my experience of living with pain might help teach me how to live with fire. I must have thought that as a person in pain, I had some sort of advantage in navigating indescribable forces. But on the morning of my first prescribed burn, under the oaks in a wild intimacy with a small piece of dirt and flame, fire begins teaching me to redefine what it means to be alive at all. I always thought of fire as that hungry monster, gobbling up everything in its path. I hadn't yet understood fire is already part of everything. Fire doesn't consume; it becomes.

The Cuban American feminist artist Ana Mendieta often used elemental materials in her work—blood, water, fire, dirt. Mendieta worked in the 1970s and 1980s, until she died in 1985 after falling out a thirty-four-story window; her husband, another artist, was thought by many to have been involved in her death. In her Silueta series Mendieta created installations in which she blended her own body into natural settings, embodying a tree, a puddle of mud, a patch of forest full of flowers, etc. In 1976 she made *Silueta en Fuego*, a film for which she dug an outline of her body in the dirt and set it on fire. I first saw the image of a woman's silhouette in the earth, burning, many years earlier in an art history course. After the fire on the path had terrified and attracted me, I recalled this image. The tree that had burned inside itself reminded me of Mendieta's art. Her work was concerned with displacement, identity, and gender; she was displaced from Cuba and her family at age twelve, and as an adult she saw her art practice as a return to a "maternal source." Her siluetas were, she wrote, "a dialogue between the landscape and the female body." Today, at the springtime burn that is miles and years away from the path, as dirt grows into flame on the land, the image of Mendieta's burning silhouette overtakes my thoughts. I wonder what it felt like for her to climb inside a hole in the ground shaped like her own body, to sculpt the soil around her hips until it took her form, then to step out of it, throw a match behind her, and light it all up.

A brief pyronatural history of women:

About a million years ago, humans figured out how to use the fire that came from the sky. The first fires they kept were in caves on the African continent. In the years before years, fire was used as protection, warmth, and an alchemical tool that rendered food safe and nutritious. The work of the hearth fire was probably done by women.

She bent over a stone hearth squatting or sitting, always bowed low, head over the flame. She must have kept it burning into night, kept the people safe, kept them fed and warm. Her neck ached and her ankle tendons grew strong. One wonders how many years were taken from her by the smoke inside her lungs.

To transport fire, to keep it alive, a person carried a smoldering coal. The spark had to be wrapped, bundled, and protected. It was tended closely at every stop along the route, sheltered from wind, rain, and thieves. It was sometimes restarted using a piece of fungus as a flint. Often only men were allowed to perform this job, and just as often they handed the flame over to the women as soon as they needed anything done with it.

Goddess of the hearth, goddess of flame, goddess of the hunt, unwilling goddess of the underworld eating pomegranate seeds like popcorn while she watches the fires of hell like TV.

A goddess took fire and she made from it islands.

A potter took fire and she made from it vessels and jewels.

The first known cremation in human history was of a woman.

The woman's people burned her body near a lake in Oceania; about forty-two thousand years later, her charred remains were dug up by an archaeologist from a colonizing country and studied. In 1992, the remains were repatriated to her family's descendants and buried in the aboriginal tradition.

In the seventh century B.C., in the temple of Vesta, goddess of the hearth and the fire within it, a flame burned that was believed to symbolize the life of the Roman Empire. The empire's fire was tended by virgins, women chosen in childhood to live as priestesses of the goddess, who in myth was both virginal and maternal. When Vesta's flame needed to be relit, the virgins rekindled it with specially rigged reflective mirrors, tilting and pivoting these instruments until the rays of the sun converged and lit the fire again. A Vestal virgin enjoyed freedoms other women didn't; she could own property and vote. She could be walking down the street and come across a person who was enslaved, and she could decide to stop and extend her chaste arm from beneath her garments and casually, almost lazily, place even one uncalloused finger on the enslaved person; and the person would then, legally, be free.

A goddess once set herself on fire. She did this to protest her repressive father, who disapproved of her marriage to a revered god. Many centuries later, her name, Sati, became synonymous with a rare and now-outlawed practice in which a widow chose—or was sometimes forced—to kill herself after her husband died. Sometimes she achieved this by being buried alive with his body; sometimes, she self-immolated on his grave.

For thousands of years fire was touted as a cure for hysteria, wandering wombs, and demonic possession of the feminine organs.

In the aboriginal language Dyirbal, which at last count was spoken by eight people in the world, there are several different categories of nouns. The category of feminine nouns consists of those related to women, fire, water, violence, and exceptional animals.

In 1600s Europe, women were allowed to be blacksmiths, but only if their husbands had been blacksmiths and were also dead. The journeywomen formed flame into metal, hammered, smelted, crafted nails with which to hold things together. They knew the names of everyone's horses. They knew heat and their bodies knew burns and heavy things.

In *The Trojan Women Setting Fire to Their Fleet,* a 1643 painting by French artist Claude Lorrain, the artist depicts a scene from classical literature in which the women of the defeated city of Troy, after wandering for years, were encouraged by a goddess to set fire to their ships and make their home where they stood, in what is now Sicily. The image shows eleven women and a baby gathered on a beach where a dozen or so tall ships are anchored; three women in a rowboat are landing on the sand near them. One woman, central, in royal blue, holds a smoking torch. At first glance, fire isn't visible in the painting; the eye is drawn to the soft blue sky, the tourmaline waters, and white smoke rising from the woman's torch. At second glance, the viewer can see dark licks of flame rising from the anchored ships. On board, figures with anguished faces wave bare arms. The luminous hues of the sky, evocative of a sunset, are revealed to be those of smoke.

In New York in 1818, a Black woman named Molly Williams was one of the only people who showed up to fight a structure fire in Lower Manhattan. Williams was reported to have pulled the fire engine—a water pump that firefighters manually hauled with ropes—by herself. She was credited as the first woman volunteer firefighter. Williams may not have volunteered out of sheer altruism: she was a formerly enslaved woman who had bought her freedom decades earlier but still worked for her former enslaver, Benjamin Aymar. Such work included tending to the needs of the volunteer fire crew Aymar was on, cooking and cleaning the firehouse. (At the time, wealthy men made up most volunteer fire crews in the city, as they had the most property at stake.) The firemen were all sick with the flu on the snowy night that Williams, who had been caring for them at the firehouse, became the only member of Oceanus Engine Company 11 to show up to a fire. She was seventy-one years old.

In Victorian England, servants rose every morning in the cold dark to silently light the chamber fires of noblemen and noblewomen. While they were there, they emptied the chamber pots. If a man lit the fires of royalty, he had a job title: fendersmith, keeper of the fireplaces and the fires within; if a woman did it, she was called a maid.

In mid-1800s America and Europe, scores of ballerinas caught fire because they danced on stages lit by gaslight, wearing costumes made of diaphanous, open-weave fabrics that moved like clouds and gathered air within their folds. Sometimes an entire corps would go up, each dancer catching the next. Sometimes the theater, too, would burn. The same types of fabric were used in trendy dress styles of the time. The writer Henry Wadsworth Longfellow's second wife, Fanny, was burned to death by her own diaphanous dress in their home. Oscar Wilde's two sisters died together in a similar fashion. It was estimated that in England alone, three thousand women died of dress fires within one year.

In 1911 in New York City, fourteen-year-old seamstresses Kate Leone and Rosaria Maltese, who were immigrants like most of their colleagues, were either burned or suffocated or jumped to their deaths at work. One hundred and twenty-one other women—the girls' coworkers at a dress factory—died along with them. Later it was found that the factory doors had been bolted shut, to prevent the workers from taking breaks. Before their deaths, Kate and Rosaria worked six days a week, labor that might earn them as much as $10.

Between 1912 and 1914, suffragettes organizing to win the vote for women in the United Kingdom conducted a direct-action campaign using arson as a tactic. During this time there were as many as one hundred incidents in which bombs and other incendiary devices were planted in government buildings and other infrastructure targets by suffragettes affiliated with the Women's Social and Political Union. One activist, a former dancer named Lilian Lenton, stated that she had a goal of burning "two buildings a week," whenever she wasn't in jail.

In the 1944 Howard Hawks film *To Have and Have Not,* nineteen-year-old Lauren Bacall holds a lit match in her right hand and eyes a mark while it burns. Her lips are closed around a cigarette that droops, ready for the flame. Her eyebrows are arched peaks; her big-shouldered suit is cut in a tight houndstooth. She looks unimpressed.

In Hiroshima, Japan, a decade after the United States dropped an atomic bomb on the city, a group of young women started a support group that was moderated by their minister, Rev. Kiyoshi Tanimoto.

The women had been severely burned during the bombing and subsequent fires. They became known as the Hiroshima Maidens, because their injuries were said to have made them unmarriageable. Reverend Tanimoto had been featured in *Hiroshima,* John Hersey's legendary *New Yorker* article and subsequent book; Tanimoto used his moment in the spotlight to start a charity that would bring the women to the United States for reconstructive surgery. Upon their arrival in the States in 1955, the Maidens received significant media attention. *Time* magazine ran "before and after" photos of Shigeko Niimoto's face, captioned *horror* (pre-surgery) and *triumph* (after). Two of the maidens also appeared in a bizarre episode of the TV reality show *This Is Your Life,* which focused not on their lives but on Tanimoto's. On the program the Maidens appeared only in silhouette, seated behind screens, "to avoid them any embarrassment," said the host, Ralph Edwards. The program also featured a surprise appearance by Captain Robert Lewis, the copilot of the plane that had dropped the bomb. On television, the airman appeared pained and stiff; Tanimoto appeared stunned. Viewers were encouraged to donate to the women's reconstructive surgery costs. Captain Lewis, looking excruciatingly uncomfortable, made the first donation, followed by the program's sponsors, cosmetics brand Hazel Bishop and Prell Shampoo ("Try some!" Edwards shilled as he brandished their $500 donation checks). The host then turned to Captain Lewis and said, "I know that our audience will be just as generous as you are, for this is the American way." One of the Hiroshima Maidens, Miyoko Matsubara, declined to travel to the United States with the others. She stayed behind alone because she "just didn't feel right" visiting the country that had done this to her. Later in life she became a peace activist and traveled widely. She never married.

In 1991 in Japan, French volcanologist Katia Krafft was immolated along with her husband in a volcano eruption. In film footage taken by the couple many years earlier, Katia, in fireproof boots, confidently took one step onto a shelf of volcanic rock, but the rock was still active lava, and flames flared up around her foot. In another home movie, she ate eggs fried by her husband on lava rocks still hot from a volcano and smiled, her eyes full of love.

In 2022 in Iran, during widespread protests against a repressive regime, women began to set fire to their headscarves as a symbol of rebellion. At one protest, a sixteen-year-old schoolgirl, Nika Shahkarami, was seen standing on an overturned trash can and waving a black headscarf that was on fire. Security forces disappeared her that night, and her body was found a week later. It was reported that the girl's mother, Nasrin Shahkarami, was pressed by the police to deny the regime's involvement; they threatened that if she didn't go on television to corroborate the official, false version of the events surrounding her daughter's death, she would be burned. Nasrin was said to have responded, "You can't burn women made of fire."

Then of course there were the witches who burned, burned, burned.

Winter, 2023, California. Extreme weather has swung its pendulum again and hard rains are falling, flooding, downing trees. Near a small town in western Sonoma County, a local journalist watches a girl, maybe seven years old, playing with fire in the rain. The girl wears a pink tie-dye hoodie and holds a little bundle of sticks, which she adds one by one to a pile of embers. She is among fifty people attending an event promoting Indigenous basketweaving, cultural fire, and the connections between them. The land they are on is owned by a Native-led nonprofit and includes a small farm, residences, educational resources, and refuge space, all based on principles of traditional ecological stewardship. At the fire pile, the journalist leans down and asks the girl what she thinks about all this. The kid looks at her mother, who nods, like, go ahead, honey, it's okay. The girl answers, I love fire because I am fire. The reporter smiles and asks, What's your name? Fire, the girl replies. A double take: I'm sorry, what did you say your name was? The kid pauses, sighs, grown-ups are so annoying, they always want to put words to everything. My name is Fire, she says.

This was not a metaphor: seabirds in flight over the inland forest. They were telling me something. They were saying it was going to rain.

I was at home. Mid-November. I didn't know what time it was; the sky was gray. I heard the unfamiliar sound of something tapping on the skylight window. Falling. Not wind or branches. Not ash or the sun's harsh heat. It was water; it was rain.

Beauteous relief, solace in motion. It fell and fell. This first rain did not pulverize or flood. It was a gentle rain: it entered with ease into the soils and bedrocks and did not harm them. Everything on Earth opened to the water.

The rain fell first on the tallest trees, who drank and waved it onward, and birds stopped their bickering to shelter and sway in the wetted canopy. It fell then on green life: leaves acted as nets to collect the rain's droplets, capturing the sweet affluence of water. In places where fire had come, the rain pooled amid ash on the ground, sought its level, nudged gently at debris and downed trees, and carried them into the rivers and byways. It fell on houses and tents, highways and mountains, and on the ocean, which didn't notice it at all.

The rain fell on the house that I lived in. I listened to a world of sounds I'd forgotten existed: murmur and thunder of the trees, drip drip of the drainpipe, louder waterfall on the wood deck beneath the section where the gutter was always clogged. For a few tempered hours, the world accepted the sky's promise. The land drank. It had rained. It had rained!

When it rained in West County paths became streams. Water gathered in the flatlands between the Mayacamas and the coastal ranges, swelling the Laguna de Santa Rosa and flooding the vineyards with new habitat for egrets and fish. When it rained the Russian River grew from a thread into a deep braid of current. In the woods behind the house where I lived, the precipitation saturated the soil and there it collided with minuscule ferns and mosses already waking up, already gone green again, dew-eyed and resilient, creating their own little forests within forests.

Max and I made like the moss and went outside and stood beneath the sky, heads up to it, mouths open and laughing. Rain fell gently upon us. It wetted my hair and my feet. My attention had been so often fixated on the mutinies of my internal systems that it now felt jarring to be conscious of the external surfaces of my body. Rain traced the watersheds and ridges, redwoods and *Quercus*. The air outside tasted so clean; it was cleaner than clean: it was water.

· 13 ·

Dirt

In the yard of the house where I lived, the rain penetrated soil that had previously been hidden by brambles and vines. Max's labor on the hillside below the flower garden had already revealed the soft, big greens of native thimbleberry and silver-backed huckleberry leaves. With water, the seed bank—that perpetual reserve of dormant kernels waiting underground for amenable conditions—began to be activated. The seeds of white-flowered stalks called Hooker's fairybells, previously shaded beneath briars, began to germinate. They'd rise by early spring. Belowground the bulbs and rhizomes of late-winter wildflowers like trillium and fetid adder's-tongue began to stir, tapping into the carbohydrates they stored last spring to power their emergence. The tropical-looking leaves of false Solomon's seal, which wouldn't unfurl until April, were already being photosynthesized into existence. Volunteer foxgloves began the long process of creating their towers of little pollen mittens, and bold oak seedlings sprouted in the shadows of their progenitors. By summer the meadow in the yard would fill with crimson clover and orange California poppies shaped like little hats.

It was still November, still fire season although with an air of slightly relaxed vigilance. The rain had come once but it hadn't come again. It was shaping up to be another record-breaking dry winter. The fire danger was still present, the winds strong. But it felt to me as though there had been a collective exhale, a regrouping. Everyone was getting more sleep. And now that it had rained, chain saws could safely be operated again without fear of sparking new fires. Today our neighborhood fire-safety group was getting together to clear defensible space.

In the WUI, where buildings intermingled with vegetation, the biggest ignition risk for houses was embers. If the wind blew a flaming ember ahead of a conflagration and it wedged itself into a roof vent or under a wooden deck, it was truly game over. Houses on fire caught other houses on fire. Fences caught trees; trees caught roofs. Residents of areas at risk for wildfire activity were advised to reduce risk and make their neighborhoods safer for firefighters by practicing home hardening, which meant mitigating a house's vulnerabilities and clearing defensible space, which included leaving a perimeter of fire-unfriendly materials around a home.

Outside the nearest Cal Fire station there was a big sign that read WHY 100 FEET? The recommendation was to have one hundred feet of defensible, vegetation-free space around the house. The infographics that Cal Fire left on our doorstep showed big, square houses on big, square lots, surrounded by empty land. In the former logging camp, creating a vegetation-free perimeter around a house was functionally impossible for those of us whose homes were beneath trees. More applicable to my interests, the hundred-foot perimeter was broken down into zones of risk: thirty feet away, you were supposed to limb up trees to six feet, to prevent fire climbing to the canopy; closest to the house, in the ember zone, they recommended no vegetation at all. A gravel or cement perimeter around the home would be ideal. In general, residents were supposed to use only certain kinds of trees in our landscaping, and to remove or replace fire-prone native vegetation such as fir, live oak, manzanita, and bay laurel. It was suggested that wooden decks might be replaced with non-wood materials and open eaves might be enclosed (a recommendation I didn't even understand). No lawn furniture made of wood, no fences made of wood, no fences or decks or furniture touching the house. No wood mulch in the garden. No less than six feet of cleared vegetation on both sides of the street, for evacuations. The streets in my neighborhood were hardly six feet wide in the first place.

Among those paying attention to such matters, increased hardening of homes and neighborhoods against fire was widely perceived as essential disaster preparedness. Heading into the megafire era—in which wildfires moved faster than cars—such adaptations would unquestionably become even more important. They were also wildly expensive. Most of my neighbors, myself included, couldn't afford to make those types of changes. For those who didn't own their homes, landlords were often absentee. Even if upgrades were doable, in Sonoma County, the amount of structures being rebuilt after previous wildfires had made it functionally impossible to obtain the services of a contractor, and building materials were getting more expensive. California's looming insurance crisis added to financial burdens for people in high-risk fire areas. Three of our neighbors had had their home insurance canceled in the past year. Too risky. Where I lived, even residents who could afford home hardening might not want to do it: the aesthetic of a fire-resilient home was in direct opposition to the woodsy, ramshackle style that gave this place its rustic charm. I had heard from old-timers in the neighborhood that decades ago the fire department would periodically go around and suggest to homeowners that they trim trees that overhung the road, in the event of an evacuation. The campaign was not popular. One neighbor who came to our workday told me she was meditating to ask the elements to stop the fires; she did not want to cut any trees. In recent years, as property had become exponentially more expensive, this aesthetic—the whole NorCal vibe—had persisted. I was no different than a land-rich ex-hippie; the untended forest was an aesthetic I'd admired since childhood, and overgrown, allegedly wild vegetation was central to my concepts of beauty and home. I would always have to struggle to unlearn the fundamental underpinning of my twentieth-century settler environmentalist values: You don't cut down trees. Ever. But lately when I looked at the woods around me, instead of a secret garden I saw a neglected one, going up in flames.

I knew by now that cutting could mean care, although I still approached the prospect of creating defensible space with trepidation. I wanted to do everything I could, of course, but I had limited funds and limited physical capacity. Besides, if a large wildfire did come through this cluttered forest, this little wooden house under big trees—on a hill filled with hundreds more acres of trees—would definitely burn down. I figured denuding my porch of its wisteria vine wasn't bound to help much.

By contrast, there was a lot that could be done about the forest floor. The hill above the house where I lived was stacked high with vegetation, much of it dead. Even the fire path was encroached on by eager bushes and upstart firs. Shaded fuel breaks—areas where hazardous ground fuels are removed while leaving taller, more fire-resilient trees—probably weren't going to stop one of these new mega-fires in the throes of a devil wind. But, especially around dense WUI neighborhoods like mine, they could help reduce risk. And so here I was, aesthetics be damned: sitting on my ass in moist dirt, pulling English ivy off native ground cover, and cutting up dry branches that my neighbors had deposited in a pile. Today my pain was bedrock, so solid it might be taken for granted.

Max and I had a new plan that we hadn't told our neighbors about yet, which was to reintroduce fire into these woods. The forest behind and between the houses in our neighborhood, which included the path where our Halloween fire had been, was owned by a private party who performed no vegetation management. Max, ever the organizer, had recently started conversations with the owners about allowing us to do work on the property. To prepare a piece of land for intentional fire required significant attention, handwork, and cooperation. It also required insurance. We'd cleared the first hurdle—liability waivers—and the neighborhood had already spent one afternoon clearing brush with hand tools. The plan was to keep going until, whether it took seasons or years, we could see the ground below the trees on the hill.

As I worked, each pull on the ivy seemed to tug inward at me, ratcheting down my arms and back into my butt, my thighs. Max and a couple of other neighbors were up on the slope, downing dead tan oaks. Others were making piles of the fuel we were clearing, to be chipped later. People felled short bay seedlings, chopped them up with loppers, and laid them flat on the ground to decay and add organic matter to the soil. Several decades' worth of dumped trash were hauled out of a seasonal streambed. Heinrich, a neighbor who worked at an apple processing plant, came and brought his chain saw, as did Kyle, a neighbor who worked doing this exact type of vegetation management. Kyle also had a wood chipper, a precious piece of equipment around these parts. Friends came, too. Davida and Lina were then beginning the process of moving up to Sonoma County from Oakland. We talked about their plans for a vegetable garden at their new spot as they ferried bundles to the chipper piles. Benjamin and Zoë had been sent to us by a mutual friend; today was the first time we met in person, though we were soon to become close. Frank stopped by, wearing no shirt and holding a can of Diet A&W. It was a party.

Benjamin and Max set to dismantling a pile of trash someone had wrenched out of a culvert. A cool thing about fuel mitigation work was that it could also be erosion and flood mitigation work; after we lopped low-lying and dead branches from trees, the branches would be laid horizontally on the hillside. This would protect the hillside from excessive water runoff during rain, benefiting both the neighborhood and the watershed.

Benja was from Venezuela and Zoë was from the United States; they had met and married in Venezuela fifteen years earlier, and they lived in Benja's family home in the rainforests near the southern border with Brazil. The collapse of Venezuela in 2015 displaced more than seven million people; water shortages eventually pushed Benja and Zoë across the border, to the Amazonian region of Roraima in Brazil. Since the pandemic they'd been living half-time in Sonoma County, to be nearer to Zoë's family. They were the kind of people who instantly felt like longtime friends; we shared many affinities. Benja was a filmmaker and Zoë a writer, and both were involved in Indigenous activism in the Amazon. As Benja piled twigs and trash, he asked in Spanish, Isn't this something the government should be doing? Yes, Max replied. Yes, it should.

My neighbor Christie and I had spent the last few weeks trying to get the county government to fix Christie's street, which had been so degraded by the elements that it had turned back into a dirt road, riddled with potholes that were a foot deep. It was impassable, and the failure meant there were fewer evacuation routes. (Because their road technically no longer existed, some neighbors also couldn't get package deliveries at their houses; Christie used our address to get garbage service.) It hadn't been the first time we'd been unsuccessful at getting help with infrastructure issues in the former logging camp; in rur-burbia, we occupied an odd civic purgatory. We didn't have local government representation except at the county level. And we weren't wealthy, like much of the nearby town. The county had stopped fixing our roads years ago. We were a high-risk area for fire; among neighbors I joked without really joking that the powers that be had already written us off as dead.

In response to Benja, I laughed.

Sorry, he grew up in a socialist country, said Zoë, faux apologizing.

At the end of the workday our labor in the brush was almost untraceable. Before and after pictures were indiscernible. But because I now knew the slope intimately, I knew how it had changed. There were spaces now between some of the dead tree stumps; the dark brown of earth was visible on the hill where before it had been the faded brown of dead green matter. The huckleberries might have fruit this year, now that they could actually see the sun. But the cleaned-up hill still looked to me newly barren. Every day when I was in the yard or the driveway, I practiced getting used to it. This is how a redwood forest should look, I told myself. This is the way we get to good fire.

After my day raking dozer lines in the nature preserve, I had realized that the more time I spent in burned places, the less I feared fire. Today, touching detrimentally unburned places, I found myself focusing more on the people alongside me than on the elements. As I worked I found I wasn't talking to the trees anymore. I was living—a verb—in a place where others—human and nonhuman—also lived. These informal neighborhood workdays and the many meetings that had led up to them were building new relationships between people and place. It was tricky and slow, like handwork. The fires had taught me this already: Whatever you did, it had to be local. Start with the house. Clear the brush and branches. Then the street. Make sure the routes out are accessible. Then turn to the forest that lurched its shadow over you, the hill on which you lived, and see where a difference might be made. The pandemic had made local actions the only kinds of actions for a while, but the fires made it clear to me that the micro was also a place where I might begin to reckon with the enormity of this moment.

Would our work alter the trajectory of the wildfire crisis? Absolutely not. Would the houses we lived in burn? As José Luis said to me, the only thing we know about fire is that *we don't know.* Did I now know the name, number of pets, and fire-related anxieties of every person on my block? I did, and they knew mine.

Staying in Sonoma County wasn't a decision Max and I ever made definitively; it was what happened, and it would keep happening until it couldn't anymore. Fire kept happening. The planet kept warming. Capitalism did its thing, ouroborosly. Max and I continued to fill the vessel of our life together: we gathered more community at the house in the woods and invested ourselves in more projects that might benefit the land on which we lived. We committed to learning more about the land's past and how those violent histories facilitated our lives here, and we would always continue to struggle with how to make it right, at scales large and small. We looked for cat eyes in redwood stumps. We began to touch places that had burned, and then to touch fire. We began to shift our careers closer to topics that addressed the crises at hand. And I planted more roses, thinking of the temporality of my existence as I pruned them. I was working to replace fear with care but in many ways the fear was still there. Despite this, and despite my resistance to hashtag optimism, the wildfires were making me more hopeful, although not in the ways one might expect. Instead of a blurry, end-of-the-rainbow fantasy of salvation, the hope I was finding was more practical. It was in these reciprocities. Living with fire was so, so messy, but fire wasn't going anywhere. Neither was I.

Andrea Bustos, the good-fire educator, put it this way: I once asked her if she ever wanted to move back to the city. She had lived in the Ecuadorian capital of Quito for a long time, and had left to work with fire in rural Indigenous communities. She told me she didn't miss the city, now that she had a connection with the land. But, she said, it didn't matter if she missed it or not. Once she built a relationship to the land, she couldn't leave. You don't light a fire and then walk away from it, do you? Once you light a fire, she said, you become responsible for that fire and that land. She offered the example of some invasive thistle that she'd seen growing in my garden. If you want to get rid of that thistle, she told me, you need to put fire on it. And the next year it will grow back. So the next year, you need to come and put fire on it again. It's a long-term commitment. What would happen if you just walked away?

After people went home and the noise of Kyle's chipper faded from my eardrums, I rallied my muscles and headed up the path. As I climbed the steep slope my pain quivered, an aftershock. Before I headed deeper into the trees I turned, looked down the hill, and beheld the forest I wrongly thought of as mine. From this part of the path I could still see the house that the papers said I possessed. All I had to lose. Some utopia: The porch crowded with muddy shoes, the potting shed with the door that wouldn't shut fully, the roof with that one mysterious leak. Around them the flowers, the Japanese maple trees, electric red in November, the sizable rhododendron bushes that had some sort of disease but every spring kept churning out saturated, moody blooms that reminded me of old British novels. This half acre was where I was learning to live beneath trees and to unlearn the idea that I was one of them, while also understanding that I had a role to play in their lives. There was no idyll here, only an unvoiced plan that was taking form as Max and I improvised. As I took in the view, I tried to think a thousand years into the future when life would certainly be harder in this place, and when what I recognized as *this place* might no longer be recognizable as such. Imagine how much careful and local attention would be needed, then, for animals and plants to stay alive. In the meantime, I was here. Disparate forces and contexts had brought me here, but at the end of the day I lived here, and so whatever happened to this place would happen to me. In a way, it was really that simple: what happened to the land would happen to me.

The morning after the neighborhood workday it rained again. Outside, the air smelled fresh: late November. Inside, on carpet, with the heater on, I did my daily stretches. I performed the movements the body therapists told me to do. Small strengthenings: a yoga position called bird dog; I was neither. I tried to seek out the sensations between fleshes, the substrata of my musculature, unseeable layers inside. In a singular instance, I made a motion and one small thing inside was shifted or touched: a direct hit. Between my hip bone and the fatty parts behind it, below the leftover organs that I had been told I couldn't feel but did, a specific place in my body was pointed at—I didn't like the phrase trigger point, it was so violent, but I had found a sensitive spot, and with my movement it had been poked. This was the goal of the stretch, but it wasn't enjoyable. Feelings surfaced, and I collapsed. My experience folded inward again. Water came out of my face, my legs corkscrewed into themselves, everything trembled and twitched. For hours after the stretch, this lone point wrecked me. There must have been memories in there.

The writer Kathy Acker, when she was dying of cancer, wrote that she felt her body physically remembered everything that had happened to it, but the physical was only one aspect of the body. In pain, the physical, the nervous, the "energetic," as Acker put it, and the emotional aspects of the body combined their experience. In doing so, they sometimes ended up re-creating the memory over and over until it didn't need an inciting incident. Pain fragmented the memory of an experience before the memory was formed; it remembered things that had not yet happened. Sometimes, too, it was difficult for me to remember what had happened—hard to explain, but also to remember. How did it start? The pain, then the garden. The garden, then the fire. Fire first, before anything. Like a dream that one retells by phrasing it in the present tense, locale, causality, and sequencing were mixed up. A damaged body was also a new body, one still to discover. The body was regenerating its cells all the time. If they were to cut me open again, would they find adhesions, rings of scar tissue and viscera that tell their own stories of what happened and when?

After morning stretches it was time for the dawn patrol. Outside, Max had already begun the work of preparing the flower garden for winter. He had shoveled compost into the beds, but detail was not his strong suit and the compost lay in uneven clumps beneath the flowers and trees. I sat and began to break up the clumps with my bare hands. Low down like that I could smell the stink of it: odorous yet somehow also rich. Compost was made up of decomposed organic matter, minerals, and the macro- and microscopic critters that like to feed on them and in the process break them down. Essentially, it was nitrogen and carbon. We made ours out of food scraps, dried fallen leaves from the oak tree, various greenery collected while doing yard work, and occasional contributions of chicken manure from a neighbor. When compost was added to a soil, it increased the organic matter content; it improved structure, hydration, aeration, and the ability of soil to hold nutrients. Compost was a free or cheap way of fertilizing anything and it was, as my dad said, a panacea for all soils. The robber barons could keep their oil, he said. Compost was the real black gold.

I shifted my seat on the garden path. Mulch pillowed beneath me. The hurt place I had activated in my morning stretch still trembled somewhere in there. I could tell it was going to be a long day of pain sapping my attention. Sometimes I thought that the healing capacity of the outdoors lay largely in the fact that the small, daily monotonies of gardening helped me to pass time.

Max once researched flame-skimmers, the crimson-orange-colored dragonflies that flew through our garden each June. They had a remarkable life cycle. In its larval, or naiad, state the dragonfly lived in the mud at the bottom of a pond or stream. It did this for three years. After all that time the flame-skimmer then emerged, using its 360-degree vision in full spectacle, to whirl and trip around the yard below the redwoods, atop calendula, through salvia and sweetgrass. It mated while in flight. During its time in the mud, the insect breathed through gills that were located in its anus. When I did my deep pelvis breathing, I sometimes tried to picture my inmost muscular web not as a flower, that terrestrial unfurling, but as gills—those dragonfly gills perhaps—buried for so long in soupy mud but still inhaling, until the flame-skimmer surfaced and found new ways of moving through the world.

Three years. As old as my pain.

I didn't know how long I sat grooming the compost. Time passed. The sun moved, or rather Earth did. In pain, the sensation of time was distorted. An hour felt like a day and a day could pass without me successfully shifting my concentration away from the peripheral nagging of my nerves. I had gotten good at doing life in such a condition, but I felt diminished when I did. I entered then into a state in which I could only describe myself as feeling like I was between everything. I had conversations I wasn't present for. I touched Max and didn't feel him. I stood next to chairs and couches, never sat. I did the things—experiences, entertainment, work—but felt as though I was fluttering around my actions, not inhabiting them. After all this time it was still difficult for me to recognize, initially, the onset of the pain state. I would think instead: I'm tired, grumpy, sleepy, sad, anything else. Eventually, hours or even days later, it caught up with me: oh, right, I'm having pain. *I'm having pain:* the vague and meaningless phrase I'd settled on to describe the experience of my body.

After I spread compost below each perennial, I shoveled some mulch on top. The mulch was wood chips, flammable, redwood, from a small tree we had removed because it was leaning dangerously over the house. In a few weeks we'd scatter cover crop seeds in the orchard—bell beans, oats, and vetch—and in the spring when they started to bloom we'd chop them down and turn the green matter and its nitrogen back under. I used my *hori hori* gardening knife to break up a large root bundle in the compost below some lupines. It didn't come apart easily. The other day Max had been turning the compost, stirring it to keep the heat well distributed, and he discovered that a nearby redwood tree had sent up roots inside the compost bin, from below. The roots were getting big, the clusters of small white threads chaotic and tough. The tree must have been growing—dining—in there for quite a while before he noticed it.

In the garden I had found that it was best to think of the passage of time in units of seasons and years, rather than hours or weeks. The extension and compression of time worked in incremental phases, but it was quicker than one might expect. This mulch would become soil within six months, less if the winter was a wet one. It was warm right now, steamy inside the pile. It was already transforming. A garden struck me as a perfect unit of measurement with which to encompass the experience of trying to stay alive. At a time when visualizing even the near present had become an act of anxiety, an avalanche of contingencies, the potential chaos of my garden's future had become more acceptable to me. I made a cut on a tree now, and it could take a year to see the result. It might rain too hard and damage the soil before I got the mulch down. More fire next year would mean more ash, drying out everything at a time when it should be growing. The native wildflowers I seeded to attract pollinators might come up in the spring, or not. Voles lurked, frost threatened, coddling moths hovered. Catastrophe was the only certainty in the garden, even if it didn't always arrive. The units of time in the plant kingdom expanded and contracted as necessary for growth or conservation; they were measured only by light, water, earth—the "elementals," as Scottish writer and outdoorswoman Nan Shepard said. The garden was a call and response. Fire, too. Action and reaction. My body's pain was a reaction. All of it was wilderness, all of it civilization, and it was no longer a place in which I located the promise of escape.

I knew there were many collapses yet to come. To inhabit these cycles of damage and renewal would require new ways of being. Whether in fire or ice, famine or flood, humans would have to cede the future tense, to learn to live an everpresentness, a constant state of reckoning with the beauty and pain of what we had done to our home, this natural body. There was no more time to ask, What are we going to do, what are we going to do?—that constant and impotent refrain of well-meaning individuals in response to systemic crisis. There was only what we had done, and what we were doing now.

I rose, using a trellis to help me, grunting. I had spread the compost. I had cleared the orchard floor of leaves and other debris. I had nothing more to do as a gardener today. For a moment I stopped trying to intervene. I found a seat on the wooden bench, a vestige of the house's previous occupant that Max had unearthed from beneath a mountain of blackberries, and I beheld the apples.

The apple trees still had leaves on them. There was no more ash. They shimmered with a lively, mature green, so different from the new colors I waited breathlessly for each spring. There was something vulnerable about an apple tree too young to grow fruit. It was a tree, like any other; its perceived value had yet to be defined or extracted. For now, the value I assigned to it consisted of the relationship I had with it: water, compost, cuts, afternoon gazes from a recumbent position.

The Wickson crab apple, which I had spent so much careful effort in pruning, had grown prolifically over the season. The tree was accepting my suggestions, following my cues, and forming into a perfect open-center form, a radiant sun cup ready to fill with fruit. For a brief blink, I saw the tree's shape as it was, not as I planned to form it. It was so beautiful, so young. Just a baby. Not yet even a baby, in deep time.

Lately I had been noticing an upswell of deep time in a lot of writing, mostly by men, about the natural world. Deep time was a phrase that, like fire season, people tended to use as a sort of shorthand, a way to signal the vastness of the universe and the smallness of humanity. Deep time referred to geologic time—the span of the entire life of Earth, not only those periods when humans have been alive on it. John McPhee once described deep time as the span of a man's arm; men themselves were a sliver of the tip of the fingernail (McPhee actually said "the king's" arm, referring to the customary British measurement of a yard, but it seemed to me unnecessary to bring the monarchy into it). As climate change had become less ignorable for the literary classes, everyone I read seemed to be getting their minds blown by the idea that time has gone on for a very long time.

I heard a creaking noise on the hillside. In the breeze one of the redwoods was rubbing against the wooden fence. Soon I'd have to carve out a bit of the fence, to give it room. Was that deep time, the life span of a redwood fence being shortened by a living redwood tree? Wasn't deep time just time? The time of a tree, its rings betraying experience; time in a forest that has burned; time to a body that is changing, the timelessness of a body, in pain.

I was the gardener of this young apple tree; it might live to be a hundred. It might never fruit, or it might burn, or die of thirst. For the redwood trees above it, time stretched into centuries, beyond the reach of a nation or a king. In the burned places, it could take a hundred years for them to grow back into a shape that I might identify as tree. I would never know their ending, though I knew it would arrive sooner than it should.

It was growing cold in the yard. The pain was surfacing. The tree was squeaking. My concentric circles of experience—the uncomfortable bench I sat on, the garden, the house behind it, the shifting soils below that were once mountains themselves, the fire that would one day come here—were blurring together in a manner that defied chronology. It felt good to let go of it. To not make a story of what was happening but instead to be inside it, as it was.

I looked up to the redwoods; they were silent once more. Can we just stand together for a minute? I asked them. Please. Please let me be next to you while it happens, let me stay here as you change.

It was wild, the fact of us. These were not metaphors: leaves that spring from sticks. Organs inside a body shifting to fill the space left by what has been cut. Muscles locked in a fight stance. Photosynthesis. Fire flowers. A woman was not a tree, a tree was not a person, but our endings were bound together. It was in this forest, orchard, garden, body that I was learning to inhabit the complexities of the space between us: woman and tree, land and person. Whether beneath the shadows of charred mountains or across the porous border of a flower bed overcome with winter weeds, we stood in relation to one another.

There passed between us a give and take of care and harm, time and light, an understanding that the structures that attempted to ascribe value to our corporeal experiences would always, necessarily fail.

The forest exhaled and my body took it in. We stood together, for a while. The green of the trees was the color of nothing else.

Acknowledgments

This book was written at various times in the territories of the Coast Miwok (Me-Wuk), Kashia Pomo, Southern Pomo, Awaswas (Quiroste, Cotoni, Uypi), Mono-Monache, Northern Paiute, Central Sierra Miwok, Southern Sierra Miwok, Coast Salish, Shawanwaki/Shawnee, and Manahoac nations. After the text was finalized, I donated a portion of the advance against royalties I received to TERA (Tribal EcoRestoration Alliance). I encourage those who are able to donate money to your local Indigenous land-stewardship and land-back projects.

I'd like to acknowledge the life and work of Jen Angel, who died while I was writing this book. Jen understood well the messiness necessary to making one's own utopia. She taught me how stories can shape movements, and how to make the best chocolate chip cookies ever. Miss you, lady.

I'm grateful to those who shared with me their deep knowledge of and respect for fire and land: José Luis Duce Aragüés, Sasha Berleman, Andrea Bustos, Thea Maria Carlson and the Monan's Rill community, Lindsay Dailey, Martin Duncan, Anayeli Guzman, Joanne Kerbavaz, Eric Muller, Amy Patten, Erica Perloff, Alex Rinkert, Margo Robbins, Ana Salgado, Maria Salinas, Stoney Timmons, Nancy Vail, and Tim Williams.

Thank you for nurturing me, West County Community Farm, Winter Sister, Lightwave Cafe, Full Bloom Flower Farm, and B-Side Flower Farm. Much gratitude for the work of California Native Plant Society, Cultural Fire Management Council, Fire Forward, the Kitchen Sisters (Davia Nelson and Nikki Silva), Landpaths, North Bay Jobs with Justice, Occidental Arts & Ecology Center, Point Reyes Bookstore, Resilience Force, TERA, and the Watershed Resource & Training Center. Thank you to my neighbors, the unstoppable

OGGG crew, and to Richard Seaman for starting the conversation about fire. To all incarcerated and volunteer firefighters.

The best: Colin Dickey, Rahawa Haile, and Sarah McCarry; Kathy Chetkovich and Jonathan Franzen; the Blood Moon Collective (RIP). Also pretty damn rad: Lian Alan, Lina Blanco, Dani Burlison, Leilani Clark, Katie Coyle, Lydia Daniller, Frank Doherty, Zoë Dutka, Kelly Gray, Latria Graham, Kathleen Cruz Gutierrez, Tessa Hulls, Lydia Kiesling, Rhys Mason, Benjamin Mast, Dani Martinez, Meaghan O'Connell, Gage Opdenbrouw, James Rickman, Erin Saul, Sara Shor, Sean Strub, Saket Soni, Davida Sotelo-Escobedo, Cynthia Strecker, Jesse Strecker, Jess Taylor, Kara Vernor, Anna Wiener, and Sasha Wright.

I'm grateful to the Hedgebrook Foundation and the Oak Spring Garden Foundation for the gifts of time, space, and food, and to my brilliant cohorts at both those residencies.

Infinite respect and gratitude to my editor, Denise Oswald, for getting it from the get-go, and my agent, Susan Golomb, for gently suggesting that this idea might make a better book than essay. Much gratitude to Andrea Monagle and the production team, to Rose Cronin-Jackson and Sarah Pannenberg, to all the sales folks . . . to Pantheon. Thank you, Jessica Zed, for the elemental tattoos. Thank you, LeMonie Hutt, for your feedback.

Thank you to my parents: Mary Kay, Orin, and Stephanie. To my siblings: Niranjan, Katie, and sweet, strong Caroline. To all the Bell Alpers, with extra flowers for Eileen. To George Levenson, whose curiosity and teaching helped form me.

To everyone everywhere organizing and taking action for a just transition away from fossil fuel economies; for the rights of women, Indigenous people, and migrants; for the beautiful and enduring lives of the ecosystems you inhabit: thank you.

Source Notes

This book is a memoir, and as such is remembered truth.

In the course of reconstructing the events of the 2020 fire season I relied on newspaper and magazine articles, academic papers, social media posts, and firsthand interviews. I am grateful for the invaluable reporting done by committed local journalists in California and Oregon, in particular photojournalists Kent Porter (*Santa Rosa Press Democrat*) and Shmuel Thaler (*Santa Cruz Sentinel*). Climate change and wildfire coverage in ProPublica, *High Country News,* and *Bay Nature* magazine were frequent go-to sources. The following is a nonexhaustive account of a few other sources I used, which readers may find of interest.

In seeking information about fire for laypeople, I found useful *Introduction to Fire in California,* 2nd ed., by David Carle (Berkeley: University of California Press, 2021); *Fire: A Brief History,* 2nd ed., by Stephen J. Pyne (Seattle: University of Washington Press, 2019); and Omer C. Stewart's *Forgotten Fires: Native Americans and the Transient Wilderness* (Norman: University of Oklahoma Press, 2009). Regarding queer theories of fire, I found fascinating Nigel Clark and Kathryn Yusoff's paper "queer fire: ecology, combustion and pyrosexual desire" (*Feminist Review* 118 [2018]).

For a brief introduction to the relationship between climate change and capitalism, try *Extinction: A Radical History* by Ashley Dawson (New York: OR Books, 2016). For audacious scholarship about living through it: Anna Lowenhaupt Tsing's *The Mushroom at the End of the World: On the Possibility of Life in Capitalist Ruins* (Princeton, NJ: Princeton University Press, 2015). On Yanomami culture and Indigenous life in the Amazon: *The Falling Sky: Words of a Yanomami Shaman* by Davi Kopenawa with Bruce Albert

(Cambridge, MA: Harvard University Press, 2013). On refuge during hard times, and as an object lesson in postcolonialist archaeology: Tsim D. Sneider's *The Archaeology of Refuge and Recourse: Coast Miwork Resilience and Indigenous Hinterlands in Colonial California* (Tucson: The University of Arizona Press, 2021).

On the topic of land management, Kat Anderson's *Tending the Wild: Native American Knowledge and the Management of California's Natural Resources* (Berkeley: University of California Press, 2005) is a go-to for specifics, and I found helpful the context provided by *Traditional Ecological Knowledge: Learning from Indigenous Practices for Environmental Sustainability* (Cambridge: Cambridge University Press, 2018), edited by Melissa K. Nelson and Dan Shilling; the contributions by Linda Hogan, Robin Wall Kimmerer, Joan McGregor, and Kyle White were of particular interest.

In learning more about Indigenous life in California, I read firsthand accounts collected in *A Gathering of Voices: The Native Peoples of the Central California Coast* (*Santa Cruz County History Journal* 5 [2002], Museum of Art and History), and several studies authored by anthropologist Kent G. Lightfoot at UC Berkeley. I found illuminating Kari Marie Norgaard's *Salmon and Acorns Feed Our People: Colonialism, Nature, and Social Action* (New Brunswick, NJ: Rutgers University Press, 2019). Benjamin Madley's *American Genocide: The United States and the California Indian Catastrophe* (New Haven, CT: Yale University Press, 2016) and Roxanne Dunbar-Ortiz's *An Indigenous Peoples' History of the United States* (Boston: Beacon Press, 2015) are important histories.

Regarding women's bodies: Silvia Federici's *Witches, Witch-Hunting, and Women* (Oakland, CA: PM Press, 2018); *The Undying* by Anne Boyer (New York: Farrar, Straus and Giroux, 2019); *Tender Points* by Amy Berkowitz (Oakland, CA: Timeless Infinite Light, 2015); Audre Lorde's *The Cancer Journals,* special ed. (San Francisco: Aunt Lute Books, 1997); *The Diary of Frida Kahlo* (New York: Abrams, 1995); and Virginia Woolf's 1930 essay "On Being Ill" (Ashfield, MA: Paris Press, 2012). I learned about incendiary fashion from Rae Nudson's wildly entertaining article "A History of Women Who Burned to Death in Flammable Dresses," illustrated by Anna Sudit (*Racked* magazine, December 19, 2017, https://www.racked.com/2017/12/19/16710276/burning-dresses-history).

In thinking about conversations between people and their gardens, I found invaluable Masanobu Fukuoka's 1978 treatise *The One-Straw Revolution* (New York: New York Review of Books, 2009) and Jamaica Kincaid's gardening essays, specifically "In History" (*Callaloo* 24, no. 2 [Spring 2001], The Johns Hopkins University Press).

Excellent books about grief and place include Cal Flyn's *Islands of Abandonment: Nature Rebounding in the Post-Human Landscape* (New York: Viking, 2022); Dionne Brand's *A Map to the Door of No Return: Notes to Belonging* (Toronto: Vintage Canada, 2011); and *Ganbare! Workshops on Dying,* by Katarzyna Boni and translated from the Polish by Mark Ordon (Rochester, NY: Open Letter, 2021). Similarly excellent is Sofia Samatar's essay "Standing at the Ruins" (*The White Review,* no. 30 [2021]). (Samatar's work also introduced me to Kari Marie Norgaard's work.)

I have jump-started many writing days with the poetry anthology *Black Nature: Four Centuries of African American Nature Poetry,* edited by Camille T. Dungy (Athens: The University of Georgia Press, 2009).

Last but not least: to identify the birds, plants, and rocks that are your neighbors, nothing beats a field guide. Some local favorites are *Field Guide to the Laguna de Santa Rosa: A Manual for Identifying Common Animals & Plants,* edited by Catherine Cumberland and illustrated by John Muir Laws (Santa Rosa, CA: Laguna de Santa Rosa Foundation, 2018), Elaine Mahaffey's *Wildflowers of the Sea Ranch: Descriptions of 214 Species with 181 Illustrations* (Chelsea, MI: Elaine A. Mahaffey, 1990), and *Roadside Geology of Northern and Central California,* 2nd ed., by David Alt and Donald W. Hyndman (Missoula: Mountain Press, 2016).

A NOTE ABOUT THE AUTHOR

MANJULA MARTIN is coauthor, with her father, Orin Martin, of *Fruit Trees for Every Garden,* which won the 2020 American Horticultural Society Book Award. Her nonfiction has appeared in *The New Yorker* and *Virginia Quarterly Review.* She edited the anthology *Scratch: Writers, Money, and the Art of Making a Living,* and she was managing editor of the National Magazine Award–winning literary journal *Zoetrope: All-Story.* She lives in western Sonoma County, California.

A NOTE ON THE TYPE

The text of this book was set in Sabon, a typeface designed by Jan Tschichold (1902–1974), the well-known German typographer. Based loosely on the original designs by Claude Garamond (ca. 1480–1561), Sabon is unique in that it was explicitly designed for hot-metal composition on both the Monotype and Linotype machines as well as for filmsetting. Designed in 1966 in Frankfurt, Sabon was named for the famous Lyon punch cutter Jacques Sabon, who is thought to have brought some of Garamond's matrices to Frankfurt.

Typeset by Scribe,
Philadelphia, Pennsylvania

Printed and bound by Berryville Graphics,
Berryville, Virginia

Designed by Cassandra J. Pappas